COMMERCIALLY IMPORTANT SEA CUCUMBERS OF THE WORLD

by

Steven W. Purcell
National Marine Science Centre
Southern Cross University
Coffs Harbour, NSW Australia

Yves Samyn
Royal Belgian Institute of Natural Sciences
Brussels, Belgium

and

Chantal Conand
Professeur Emérite, Laboratoire Ecologie Marine
Université de La Réunion
Saint-Denis, France

FOOD AND AGRICULTURE ORGANIZATION OF THE UNITED NATIONS

Rome, 2012

ISBN 978-92-5-106719-2

Sea cucumbers of commercial importance are distributed globally, with most species harvested in multiple locations. Biological information is available for many species but limited for others, making resource assessments and trade data potentially misrepresented. Species within some taxonomic groups can look similar, both in the field and once processed into a dried product, so identification guides should allow scientists, traders and trade officials to understand the features of each species that easily distinguish them from other ones.

A number of identification guides and taxonomic accounts of sea cucumbers exist, but none has provided global coverage of all commonly exploited species with photographs of live and dried animals. In addition, some use technical language and characters not easily understood by people with limited taxonomic training.

International fora dealing with the conservation of sea cucumbers (i.e. FAO Workshop on Advances in Sea Cucumber Aquaculture and Management – Lovatelli *et al.*, 2004; CITES[1] Technical Workshop on the Conservation of Sea Cucumbers in the Families Holothuriidae and Stichopodidae – Bruckner, 2006; CITES Animals Committee and Fourteenth Conference of the Parties – Toral-Granda, 2007; IUCN Red List Workshop for Sea Cucumbers, Colombia 2010) have underscored a current limitation in available tools for identifying processed animals traded internationally. For example, although *Isostichopus fuscus* was included in CITES Appendix III by Ecuador in October 2003, conservation of this species has lagged, in part, because custom and border control agents have had little information by which to distinguish the dried animals from other similar species.

Consequently, the Food and Agriculture Organization of the United Nations (FAO) coordinated a project to prepare this global identification guidebook. A scientific steering committee for the project, with the help of numerous field experts, compiled a list of commercially exploited sea cucumber species (see Purcell, 2010). Information on biology, diagnostic features, exploitation and distribution of these species were then solicited from taxonomic experts and field biologists globally. Some species were excluded from this guidebook because they are seldomly exploited, or are exploited in minor quantities, or were poorly known to science, or photographs were completely lacking. Three other species exploited for the aquarium trade were also intentionally omitted because the focus of this guidebook is on species exploited for human consumption. Finally, recent genetic work funded by FAO (Uthicke, Byrne and Conand, 2010) enabled the deletion of two synonymic species: *Bohadschia bivittata* and *B. similis*.

In addition to various technical papers, several taxonomic guides were especially relied upon for technical descriptions of the live animals and their calcareous ossicles, and accounts of their distributions: Clark and Rowe (1971), Féral and Cherbonnier (1986), Massin (1999), Samyn (2003), Samyn, VandenSpiegel and Massin (2006), Solis-Marín *et al.* (2009).

Claude Massin and Gustav Paulay also provided valuable comments on the species distributions. The distribution maps in this book are based upon published accounts and personal communications and are certain to be incomplete for some regions (e.g. Southeast Asia) due to a lack of available and reliable accounts.

---

[1] CITES: Convention on International Trade in Endangered Species of Wild Fauna and Flora

ABSTRACT

Sea cucumbers are harvested and traded in more than 70 countries worldwide. They are exploited in industrialized, semi-industrialized, and artisanal (small-scale) fisheries in polar regions, temperate zones and throughout the tropics. In some fisheries, more than 20 species can be exploited by fishers and should be distinguished from each other by fishery officers and scientists. The processed (cooked and dried) animals, often called bêche-de-mer or trepang, are exported mostly to Asian markets and need to be distinguished to species level by customs and trade officers. This book is intended as an identification tool for fishery managers, scientists, trade officers and industry workers to distinguish various species exploited and traded worldwide.

This book provides identification information on 58 species of sea cucumbers that are commonly exploited around the world. There are many other species that are exploited either in a small number of localities or in relatively small quantities, which are not presented. Species in some regions with active fisheries are also not represented due to limited information available (e.g. Mediterranean species). The accounts are based on more than 170 reports and research articles and by comments and reviews by taxonomists and field workers.

Two-page identification sheets provide sufficient information to allow readers to distinguish each species from other similar species, both in the live and processed (dried) forms. Where available, the following information for each species has been included: nomenclature together with FAO names and known common names used in different countries and regions; scientific illustrations of the body and ossicles; descriptions of ossicles present in different body parts; a colour photograph of live and dried specimens; basic information on size, habitat, biology, fisheries, human consumption, market value and trade; geographic distribution maps. The volume is fully indexed and contains an introduction, a glossary, and a dedicated bibliography. Readers are encouraged to base their identifications on a combination of morphological features, samples of ossicles from different body parts and information on what habitat and locality the species was found.

Purcell, S.W., Samyn, Y. & Conand, C.

Commercially important sea cucumbers of the world.

FAO Species Catalogue for Fishery Purposes. No. 6. Rome, FAO. 2012. 150 pp. 30 colour plates.

Project manager: J. Fischer (FAO, Rome).

Technical editors: N. De Angelis (FAO, Rome) and A. Lovatelli (FAO, Rome).

Scientific reviser: N. De Angelis (FAO, Rome).

Scientific illustrator: E. D'Antoni (FAO, Rome).

Desktop publisher: E. Luchetti (FAO Consultant, Rome).

Cover: E. D'Antoni (FAO, Rome).

ACKNOWLEDGEMENTS

The authors thank staff of the FAO Fisheries and Aquaculture Department, which supported the production of this field guidebook. We are indebted to Alessandro Lovatelli who provided vision into realizing this project and providing guidance to its preparation. We thank Nicoletta De Angelis for her valuable support and scientific revision, and Emanuela D'Antoni for the beautiful and accurate scientific illustrations. Fabio Carocci prepared the distribution maps, and Enzo Luchetti did the graphic design and layout creation.

The preparation of this document was possible thanks to funds provided to FAO by the Government of Japan through the Trust Fund Project GCP/INT/987/JPN on "CITES and commercially-exploited aquatic species, including the evaluation of listing proposals" coordinated by Johanne Fischer.

The scientific committee for this project provided guidance to the layout and information fields for the identification sheets. It comprised Chantal Conand, Steven Purcell, Sven Uthicke, Jean-François Hamel and Annie Mercier. We extend special appreciation to Gustav Paulay, Claude Massin and Alex Kerr for the review of the taxonomy of the concerned species and for checking distribution maps.

A special thank you and recognition goes to Veronica Toral-Granda for the preliminary work that enabled the gathering of the initial set of information and data used in the production of this guidebook. Recent retail and wholesale prices of bêche-de-mer in Hong Kong and Guangzhou were obtained by S. Purcell through PARDI project 2010/004, funded by the Australian Centre for International Agricultural Research.

This book also benefited from the collaboration of sea cucumber scientists worldwide who provided the information included in the species fact sheets, without their information and support this book would not have been possible: Mohammed I. Ahmed, Jun Akamine, Irma Alfonso, Khalfan M. Alrashdi, Juan José Alvarado, Luis Amaro-Rojas, Riaz Aumeeruddy, Mark Baine, José Enrique Barraza, Milena Benavides-Serrato, Maria Byrne, Chen Jianxin, Peter Collin, Poh Sze Choo, Niki Davey, Hampus Eriksson, Vanessa Francisco, Ruth Gamboa, Beni Giraspy, Elena Gudimova, Chita Guisado, Héctor Guzmán, Jean-Francois Hamel, Carlos Roberto Hasbun, Alex Hearn, Philip Heath, María Dinorah Herrero-Pérezrul, Ivy G. Walsalam, Daniel Baskar James, Alex Kerr, Jeff Kinch, Ambithimaru Laxminarayana, Annie Mercier, Bo Meredith, Maria Del Mar Otero-Villanueva, Gustav Paulay, Philip Polon, Erika Paola Ortiz, Mary A. Sewell, Timothy Skewes, Francisco A. Solís-Marín, Veronica Toral-Granda, Akram Tehranifard, Sven Uthicke, Didier VandenSpiegel and Erin Wylie.

TABLE OF CONTENTS

Aspidochirotida: Holothuriidae

Aspidochirotida: Stichopodidae

The use of sea cucumbers as a food item and a commodity began in China about 1 000 years ago, which encouraged the development of capture fisheries in the region. However, the rising demand of the markets in Asia led to the depletion of local sea cucumber populations and prompted Asian traders to solicit sea cucumbers from locations further afield (Conand, 2004, 2005b; Bruckner, 2006; Toral-Granda, Lovatelli and Vasconcellos, 2008; Purcell, 2010). Currently, sea cucumber fishing occurs all over the world with some populations reportedly over-harvested (Lovatelli *et al.*, 2004; Bruckner, 2005b; Uthicke and Conand, 2005; Conand and Muthiga, 2007; Toral-Granda, Lovatelli and Vasconcellos, 2008).

Most tropical fisheries are multispecific and at an artisanal scale or for subsistence use. In some cases, fishing evolved to target many low-value species after stocks of the more valuable species were depleted. In temperate regions, fishing commonly focuses on one species harvested with industrial fishing methods (Hamel and Mercier, 2008). The vast majority of species are harvested for the 'bêche-de-mer' or 'trepang' market (e.g. *Actinopyga mauritiana, Holothuria scabra, Thelenota ananas*), although some species are also consumed cooked, pickled or raw (e.g. *Apostichopus japonicus, Cucumaria frondosa, Parastichopus californicus*). Some domestic markets also demand the pickled intestines and gonads, while some commercial products have sea cucumber by-products (e.g. "gamat" oil from *Stichopus horrens*) and others are included in the aquarium trade. Generally, sea cucumber harvesting is for export, with little domestic use, and the market is largely driven by oriental entrepreneurs who set the price for the sale (Conand, 2008; Kinch *et al.*, 2008; Toral-Granda, 2008).

Since the 1980s, sea cucumber harvesting has boomed but many stocks have collapsed. Concomitantly, fishers in more and more countries are exploiting more species in an attempt to meet the strong demand in Asian markets. Towards the end of the "boom" part of a "boom-and-bust" fishing cycle, populations of some species had been reduced to such low levels that there was little capacity for natural recovery and replenishment, leading to their economic and ecological extinction.

Sea cucumbers belong to the class Holothuroidea and so are also referred to as holothurians. The majority of species harvested commercially belong to the order Aspidochirotida, specifically to the families Holothuriidae and Stichopodidae, and are mostly tropical. A few species belonging to the order Dendrochirotida, family Cucumariidae, are also fished commercially. Species in the orders Apodida, Dactylochirotida, Elasipodida and Molpadida are mostly not fished commercially and are not presented in this field guidebook. Conand (2006) recognized about 40 species of sea cucumber under commercial harvest, while Toral-Granda, Lovatelli and Vasconcellos (2008) listed at least 47 species. Later, Purcell (2010) lists 66 species that are currently exploited commonly in various regions of the world. The chronological increase in the number of reported exploited species echoes the pervasive problem of serial depletion of high-value species, leading to exploitation of new species.

The taxonomy of some groups of sea cucumbers is complex, even for taxonomic experts, and has stimulated much research in recent years. Uthicke, Byrne and Conand (2010) genetically analysed the relationships among many commercial species, shedding new light on a few of them. However, once processed, some sea cucumbers can be difficult to identify to species level, creating a problem for trade officials. This has been identified as a bottleneck when attempting to implement conservation tools in the international trade (e.g. a CITES listing) and has led to the development of illegal, unreported and unregulated (IUU) trade.

This book presents a summary guide to identifying 58 sea cucumber species exploited for human consumption through photographs of the live and processed animals, morphological descriptions, biological and ecological information, and illustrations of the calcareous ossicles ('spicules') found in various body tissues. The shape of ossicles differs among species and may be used to distinguish species in trade (e.g. *Isostichopus fuscus* [Toral-Granda, 2005]). This book also summarizes information on current fisheries and management measures of each species, although the regulatory measures currently in place may be insufficient and in need of improvement (Purcell, 2010).

GENERAL REMARKS

External morphology of sea cucumbers

Sea cucumbers have an orally-aborally (longitudinally) elongated body (Figure 1). The pentamerous symmetry is sometimes recognizable by the presence of 5 meridional **ambulacra** bearing **podia**. Sea cucumbers live on the substrate of the sea floor with their ventral surface (or **trivium**). This creeping sole bears the locomotory podia, while on the dorsal surface (or **bivium**), the podia are often represented by papillae. Consequently, a secondary bilateral symmetry is evident.

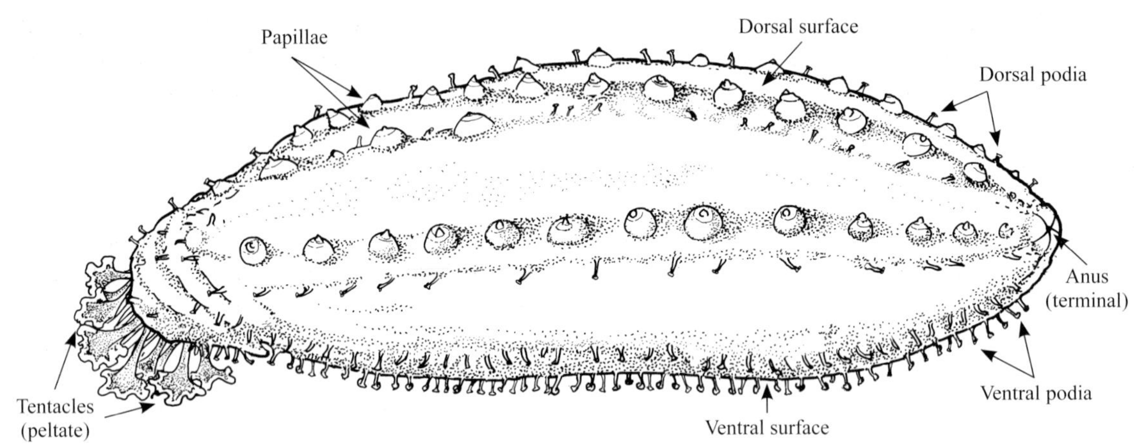

Figure 1 Main external anatomical features of a sea cucumber

The mouth, at the anterior end, has **tentacles** (Figure 1), which the animal extends to acquire food (mainly particulate organic matter). The **anus** is at the posterior end of the animal. Tentacles are buccal podia containing extensions from the water vascular system. Their number varies between 10 and 30, generally being a multiple of 5. In the **Aspidochirotida** all tentacles are of the same size, but in the **Dendrochirotida** tentacles can be of differing size. The shape of the tentacles differs among the various taxonomic orders and is used as a key character (Figure 2). In the Dendrochirotida, they are **dendritic** (branching in an arborescent manner) and can reach a large size when extended. The Aspidochirotida have **peltate** tentacles, each with a central stalk and a little branching disc. Sea cucumber tentacles are very retractile, particularly in the Dendrochirotida, which have an **introvert** where the tentacles insert. The tentacles and the introvert can be contracted into the inside of the animal by 5 retractor muscles.

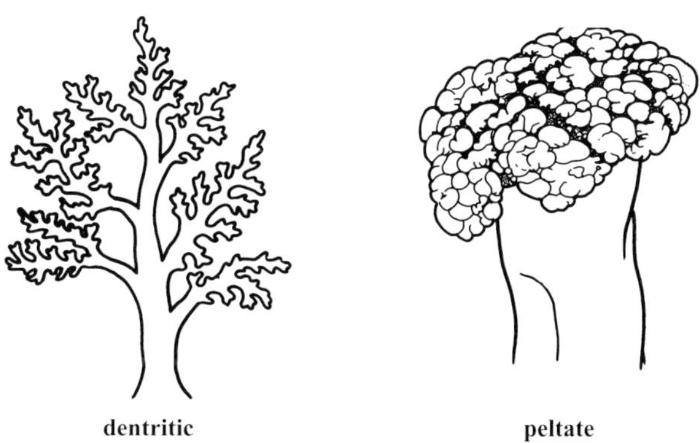

Figure 2 Basic types of tentacles

The body surface is thick, slimy in many species and bears wart-like, conical or fleshy **papillae** (Figure 1). **Podia** appear on the body wall and typically have the form of locomotory tube feet (Figure 1): hollow tubular projections terminating in a flat disc, which allows the podium to adhere to the substratum during locomotion. Epidermal cells produce adhesive secretions. Internally, the disc is supported by a large skeletal ossicle. Podia can also have the shape of papillae. The tube feet are rarely arranged in 5 regular rows, but generally they lose the discs on the dorsal surface and spread into the interradial areas. The **anus** may be encircled by small papillae or heavily calcified papillae called **anal teeth**. The coloration varies between species and sometimes also between individuals of the same species. The ventral surface is often lighter in colour than the dorsal surface.

Body wall

The body wall is thin in Apodida and Molpadida, but thicker in the other orders, particularly in the Aspidochirotida. It constitutes the part of the body that is processed for human consumption and, therefore, commercial species are characterized by a thick body wall. Its structure consists of a thin cuticle over the epidermis and a thick dermis underneath. The dermis is composed of connective tissue, enclosing the endoskeletal ossicles or 'spicules' (see next section). Below the dermis, a layer of circular muscles forms a cylinder, generally interrupted by 5 longitudinal muscle bands situated in the radial positions.

Ossicles

Also called spicules, or deposits, ossicles (Figure 3) are characteristic of sea cucumbers and of primary importance for identification. They are mostly of microscopic size. There is a wide variety of simple to complex shapes. **Rods** can be simple or branching, smooth, warty, or spiny, or can bear knobs only at their ends. They can also have a characteristic C- or S-shape. Fenestrated **plates** also come in various shapes. **Buttons** are oval ossicles, perforated with a varying number of holes arranged in 2 or more rows. **Tables** are more complicated; they appear as a perforated disc, bearing an erect **spire** (or tower) composed of pillars that can unite to form cross-beams or bridges and that terminate in a crown and show many variations according to the arrangement of its constituents. **Rosettes** are short rods subdivided into short branches. **Anchors** are peculiar of the family Synaptidae (order Apodida). They are oriented in the body wall, so that they support the attachment to the substrate during crawling, in the absence of podia. They are attached to an accompanying perforated plate, the anchor plate. Miliary bodies (**grains**) are very tiny ossicles found in some Stichopodidae. Apart from the body wall, ossicles are found in the tentacles, the podia, and often also in the internal organs. Their developmental stages can differ from the definitive shapes in the adults and thus can make species identification difficult.

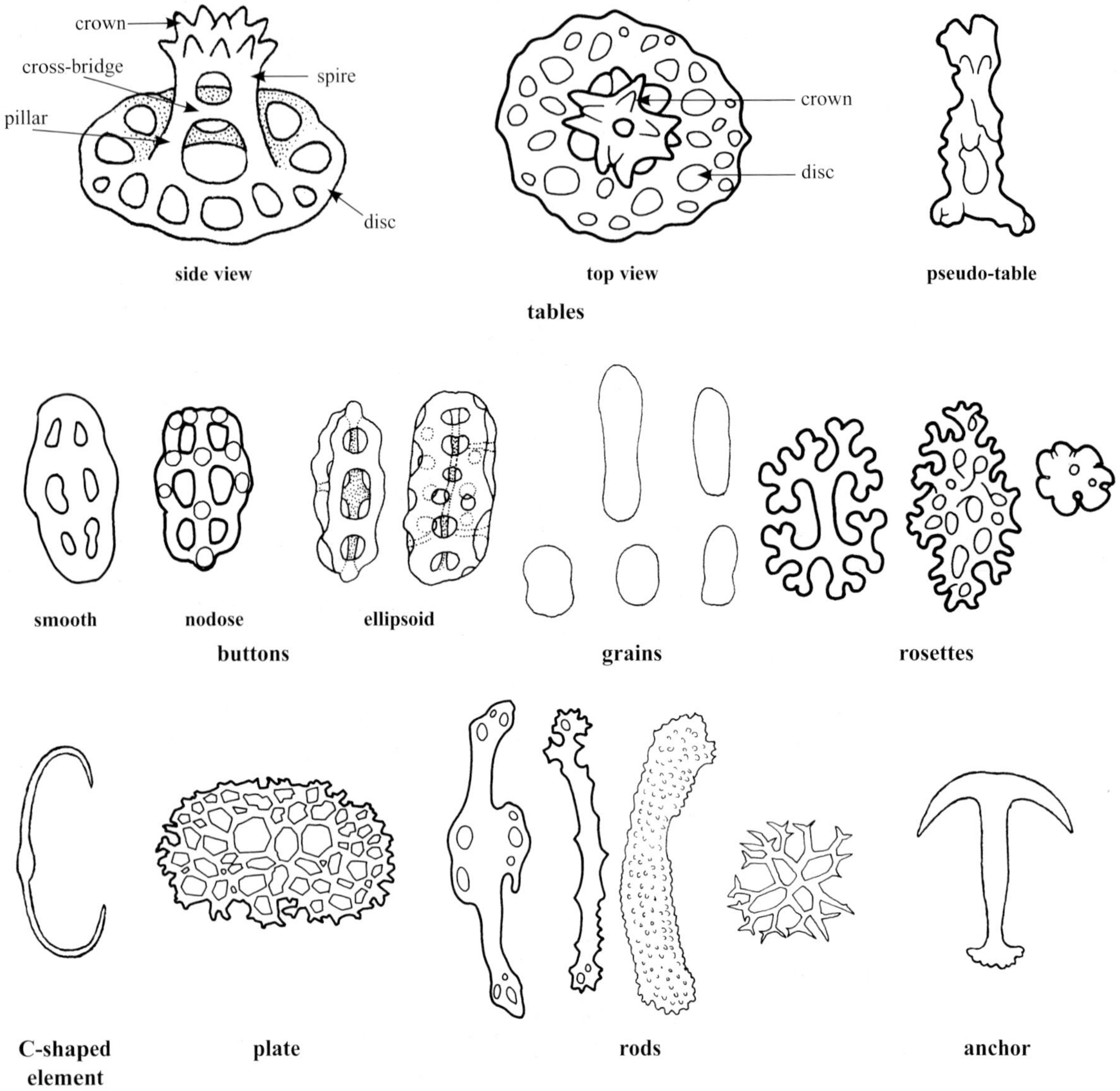

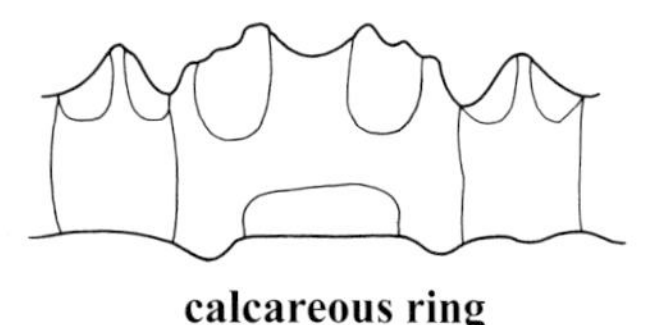

Figure 3 Basic types of ossicles

Calcareous ring

A ring of usually 10 calcified plates encircles the pharynx. It is composed of alternating larger radial plates, opposite to the ambulacra and smaller interradial plates. The plates may be simple or composed of smaller pieces. Longitudinal muscles attach to the radial plates.

Digestive system and connected organs

The gut is composed of a pharynx, an esophagus, a stomach, all of which are short structures, and a very long intestine (Figure 4). The intestine consists of 3 portions, a descending, an ascending and finally a descending loop that connects to both the rectum and the **cloaca** opening outwards through the anus. Where present, **respiratory trees** are connected to the cloaca. The oxygenated water enters the body by these water lungs, which are found in all orders except the Apodida. **Cuvierian tubules**, present in several species of Aspidochirotida, are generally considered defensive structures. They are sticky tubules attached to the base of the respiratory trees and can be expelled in some *Holothuria* and *Bohadschia* species through the cloaca towards the source of irritation.

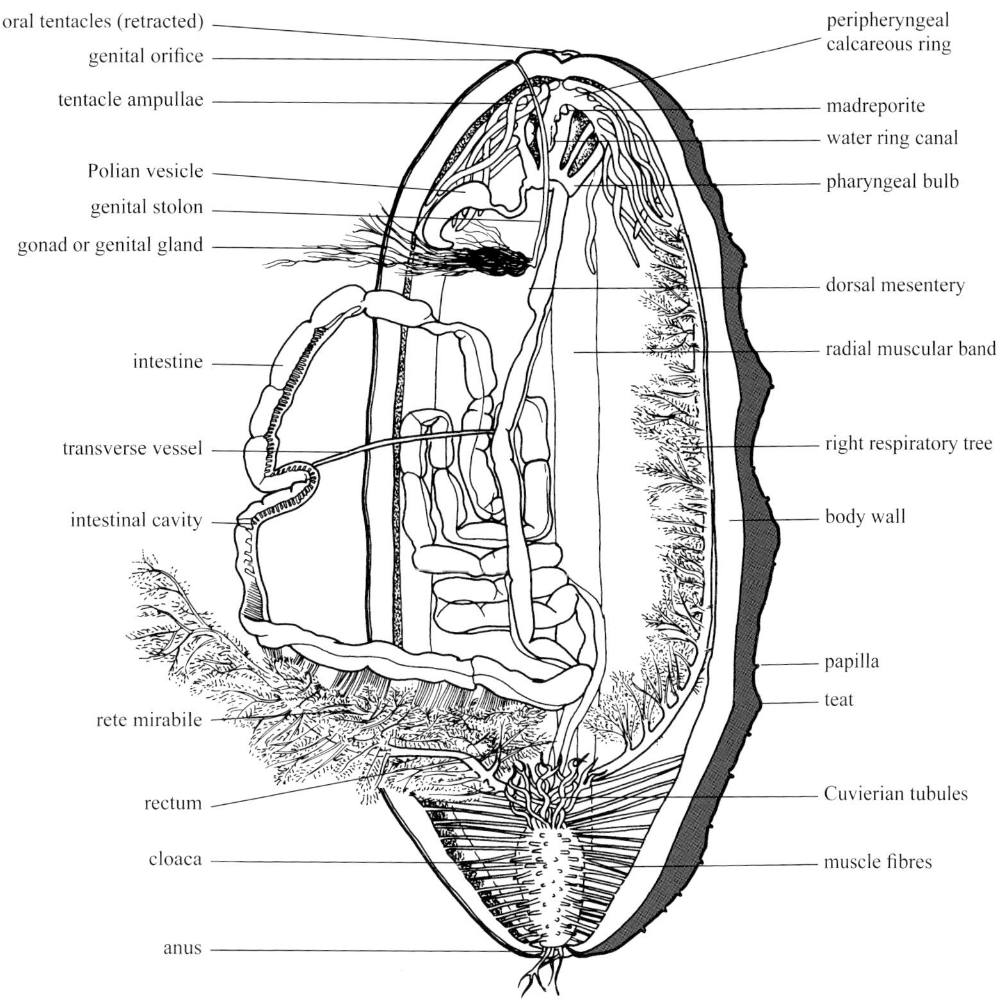

**Figure 4 Anatomy of the aspidochirotid sea cucumber *Holothuria whitmaei* Bell, 1887
(after Conand, 1989)**

Reproductive system

In contrast to other echinoderms, the reproductive system of holothurians consists of a single **gonad** or genital gland (Figure 4). The gonad is situated dorsally and in the Aspidochirotida composed of either 2 tufts of tubules (Stichopodidae), or only 1 tuft (Holothuriidae). The sexes are generally separated and show little dimorphism unless in the period of maturing. The gonad is attached to the dorsal mesentery through which the **gonoduct** or genital stolon opening passes, leading to the outside by the **gonopore** (genital orifice) or a genital papilla. In most species, the mature gametes are freely released into the seawater. The spawning behaviour, observed in many Aspidochirotida species, involves an upright posture of males and females followed by a swaying back and forth, while the gametes are being released.

Water vascular system, perivisceral coelom, and hemal system

The **water vascular system** (Figure 4) is a coelomic space bordered by a mesothelium. It consists of the lumen of the buccal tentacles and the podia, a **water ring** around the esophagus, the **radial canals**, the **madreporic canal** and the **Polian vesicles**. The perivisceral coelom is a large cavity containing watery proteinaceous coelomic fluid and different forms of cells (coelomocytes).

The haemal system is well developed and composed of large haemal vessels along the gut, sinus and lacunae. The haemal vessels associated with the gut can form a complex meshwork with the left respiratory tree, the **rete mirabile**, suggesting different functions of nutrient and gas transfers.

Habitat and biology

Holothurians are found throughout all oceans and seas, at all latitudes, from the shore down to abyssal plains. The adult stages are benthic (living on the sea bottom); some species live on hard substrates, rocks, coral reefs, or as epizoites on plants or invertebrates. Most of the species inhabit soft bottoms, on the sediment surface or buried in the sediment. Among the commercial coastal holothurians, the Aspidochirotida are predominant in the tropics, while the Dendrochirotida are more common in temperate regions. Sea cucumbers within the order Aspidochirotida have **planktotrophic** larvae, i.e. that feed on microalgae in the water column during the dispersive larval phase. Within the order Dendrochirotida, the larvae of sea cucumbers described in this book are **lecithotrophic**, i.e. the dispersive larvae feed on a lipid yolk rather than on microalgae in the water column.

Fisheries

Holothurians have been harvested commercially for at least a thousand years, occasionally for the raw body wall or viscera, but mostly in order to be processed into a dry product called **bêche-de-mer**, **trepang**, or hai-san, which is considered a delicacy and a medicinal food by Chinese and other Asian peoples. Harvesting in the tropics is usually done by hand, while wading in shallow waters, or gleaning, at low tide or by free-diving from small boats, although SCUBA and hookah have increasingly been used.

Common processing techniques

The Asian markets are now accepting new product forms of sea cucumbers, such as semi-dried vacuum packed, frozen whole or as separate body parts. Processing methods to achieve the dried form (bêche-de-mer) vary depending on the species, the final product to be achieved and the market to which the product will be sold. The Secretariat of the Pacific Community (SPC, formerly the South Pacific Commission) and the National Fisheries Authority (NFA) of Papua New Guinea have summarized the most common methods for tropical species, as reproduced below:

Method 1
Boil sea cucumber for a short time (2–5 minutes) until it swells; remove gut and body content by gently squeezing the body or by making, if necessary, a very small cut in the mouth. Put back in boiling water until rubbery and hard. Bury in a sandpit for 12–18 hours; upon retrieval, rub the outer part of the body to remove decomposed parts. Boil a third time in clean water. Drain and dry in hot air. Do not smoke. Leave to dry in the sun (from four days to two weeks depending on moisture content).

Method 2
Boil sea cucumber for a short time (2–5 minutes) until it swells; remove gut and body content by gently squeezing the body or by making, if necessary, a very small cut in the mouth. Put back in boiling water until rubbery and hard. Wash in seawater. Boil a third time in clean water. Drain and dry in hot air or with smoke. Leave to dry in the sun (from four days to two weeks depending on moisture content).

Method 3
For animals with very thick body wall, boil until it swells (may take up to 10 minutes). Slit upper dorsal side about 3 cm from each end, and remove body contents. Do not remove the five longitudinal string muscles. Wash in seawater. Boil again in clean water until hard and rubbery. Remove any remaining guts and other body contents. Place a stick across the slit to keep it open, and hot air or smoke for 12–48 hours. Sun dry for one to two days with the slit downwards. Remove sticks and tie with string or vines. Leave to dry in the sun (from four days to two weeks depending on moisture content). Remove string/vines before packing.

Method 4
For animals with very thick body wall, boil until it swells (may take up to 10 minutes). Slit ventral side about 3 cm from each end, and remove body contents. Do not remove the five longitudinal string muscles. Wash in seawater. Boil again in clean water until hard and rubbery. Remove any remaining guts. Place a stick across the slit to keep it open, and hot air or smoke for 12–48 hours. Sun dry for 1–2 days with the slit downwards. Remove sticks and tie with string or vines. Leave to dry in the sun (from 4 days to two weeks depending on moisture content). Remove string/vines before packing.

Method 5
Temperate species (*Cucumaria frondosa*, *C. japonica*, *Parastichopus californicus* and *P. parvimensis*) are also consumed raw, quick frozen or canned. The processing technique varies among countries and regions and the final product, which may be muscle strips, aquapharyngeal bulbs (called 'flowers'), gut, gonads and respiratory trees. For detailed information on individual species, refer to Hamel and Mercier (2008). These species are normally harvested and processed industrially.

Preparation of ossicles

As in other echinoderms, species identification of sea cucumbers is aided by the examination of the skeletal elements (ossicles) found in various parts of the body. The calcareous ossicles, which are hidden in the body wall (mainly in the dermis tissue), papillae, podia and tentacles are, in the species within this book, mostly just one-twentieth to one-tenth of a mm in length. They are embedded in soft tissues, but can be dissected out of the live, dried or preserved animals and isolated by the following method:

1. Small pieces (e.g. a few square mm) of tissue are removed with a scalpel from dorsal body wall and ventral body wall, as well as the tentacles and podia, and each placed into separate small vials.

2. A small volume (e.g. 0.5 ml) of sodium hypochlorite (concentrated household bleach), or sodium hydroxide, is then added to each vial in order to dissolve the organic tissue away from the calcareous ossicles. The soft tissue will be dissolved/digested in 20–30 minutes, leaving the hard ossicles to fall to the bottom of the vial.

3. After decanting, or pipetting, out the bleach, the ossicles are washed 5 times in distilled water. This step can be achieved by sucking the liquid out of the vial with a pipette, taking much care to rinse the pipette in fresh water each time so as not to contaminate a sample with the ossicles from another.

4. The ossicles can then be rinsed in alcohol and placed onto a microscope slide with a drop of a mountant (Euparal medium). They can also be put on a scanning electron microscope (SEM) stub.

5. After processing, the ossicles can be observed either on permanent slides with a light microscope, or prepared for a scanning electron microscope.

Preservation of whole animals

Whole animals can be preserved to allow a voucher specimen for taxonomic identification or for having body tissues from which to take samples for ossicles or for biological investigations. Readers may consult Lovatelli *et al.* (2004), Samyn *et al.* (2004) or Samyn, VandenSpiegel and Massin (2006a) for a comprehensive protocol for the preservation of sea cucumbers. Below is a summary account of key steps in this process, adapted to modern procedures that allow both molecular characterization of the concerned species, but also the long-term preservation of the voucher specimen.

1. Gain authorization or a permit for collecting and exporting the samples.

2. Take a piece of tissue ($1/2$ cm^2) and preserve it in 100% ethanol for molecular characterization.

3. Relax and anaesthetize the whole live animals in a solution of 5% magnesium chloride ($MgCl_2$) or magnesium sulphate ($MgSO_4$; also called Epsom salt) in seawater. The sea cucumbers will relax their tentacles out of their mouth, the podia (tube feet) will extend from the body, and the anaesthetization reduces the incidence of the animals eviscerating their organs.

4. Fix the anaesthetized animal in a solution of 10% formalin with adequate buffer (e.g. a couple g of sodium bicarbonate per litre of solution or some calcium carbonate). For large animals, inject some buffered formalin into the coelomic cavity of the animal. Leave for one day. Unbuffered solutions can dissolve the ossicles of the animals.

5. Exchange the fixative solution with 70–80% buffered alcohol and leave in this solution for one day. Discard this alcohol and replace with fresh 70% buffered alcohol. The animal is then preserved in the alcohol and can be left in this solution, with a label, for future reference. The waterproof label should at least have, in pencil, the collector's name, date, depth, location or GPS coordinates, sample code and the substrate from which the specimen was collected.

GLOSSARY OF TECHNICAL TERMS

Anal teeth: radial extremely calcified papillae encircling the anus, appearing tooth-like.

Bêche-de-mer: widely used term for the processed product of sea cucumbers (see also trepang).

Bivium: the dorsal part of the body in the pentaradiate symmetry, with 2 radii and 3 interradii.

Calcareous ring: internal collar of plates, generally 10, surrounding the pharynx.

Cloaca: anal cavity where the intestine ends.

Cuvierian tubules: threads becoming sticky when ejected out of the anus and used as a defence mechanism.

Dendritic: branching in an arborescent manner; used to describe the shape of the tentacles in Dendrochirotida that are used for suspension feeding.

Digitations: finger-like structures, used as descriptive term for the shape of tentacles.

Dorsal: upper surface of the animal.

Egg: a fertilized oocyte.

Fenestrated: having small window-like openings or holes.

Fission product: half sea cucumbers, after the animal has divided in two, in the process of rebuilding new organs.

Hookah: equipment allowing divers to breathe compressed air from a tube attached to a compressor onboard a boat.

Interradii (or interambulacra): in the pentaradiate symmetry, the 5 areas between the rows of podia or papillae.

Juvenile: the young post-metamorphic (post-larval) animal, before reaching sexual maturity.

Lateral: at the side of the animal.

Lead bomb: a heavy weight, such as a ball of lead, with a barbed shaft attached at the bottom and a string attached at the top. The weight is lowered onto sea cucumbers in deep water to spear the animals and pull them to the surface.

Lecithotrophic: development in which the larva feeds on a lipid yolk remaining from the egg, rather than eating microalgae in the water column (c.f. Planktotrophic).

Marine protected areas (MPAs): several definitions for MPAs exist; a generally accepted one is from the International Union for Conservation of Nature (IUCN): '… any area of the intertidal or subtidal terrain, together with its overlying water and associated flora, fauna, historical and cultural features, which has been reserved by law or other effective means to protect part or all of the enclosed environment'. Thus, an MPA is not necessarily completely closed to commercial or recreational fishing.

Moratorium (or moratoria [plural]): a general ban on fishing, often for 1 or many years.

Nodose (or nodulose): having knobs or bumps on the surface.

No-take zones (NTZs): sections of intertidal or subtidal terrain and overlying water delineated and legislated where no fishing or collection of certain species or groups of animals or plants can occur for a defined period. Often, an NTZ may be a special zone within an MPA.

Oocyte: a female gamete before it is fertilized and becomes an egg.

Ossicles: or "spicules", are microscopic carbonate skeleton particles in the body wall, tentacles, podia, papillae, and other body parts, useful for species identification; they come in various shapes.

Papillae: conical lumps or small fleshy extensions on the surface of the body wall.

Peltate: describing a structure that is circular or lobed with a stalk in the middle; used to describe the shape of the end of tentacles in Aspidochirotida that are used for deposit feeding on the sea floor.

Pentamerous: having 5 radiating parts, resulting in a pentaradiate symmetry.

Planktotrophic: development in which the larva feeds on microalgae (or other plankton) in the water column.

Podia (or tube feet): tiny water-filled tubes, terminating in a disc, used for locomotion.

Posterior: at the rear end of the structure or animal.

Protuberance: a part of the body that protrudes from the main part of the body, e.g. a knob, bulge or fleshy spine.

Radii (or ambulacrae): in the pentaradiate symmetry, the 5 areas with podia or papillae.

Respiratory tree: arborescent organ (1 pair), opening in the cloaca, which fills with water to enable the animals to respire.

Restocking: the act of rebuilding stocks of spawning adults in wild populations, for example by releasing hatchery-produced juveniles or adults to a depleted population.

Size at maturity: the length, or weight, at which most animals first possess gonads with oocytes or spermatozoa.

Spawners: reproductively mature animals in a population.

Subdorsal: appearing near, but not quite on, the very upper surface of the animal; half-way between terminal and dorsal; mostly used here to indicate the position of the anus.

Teats: large papillae at the border of the ventral surface of the animal.

Tegument: the outer tissues of the animal, including the cuticle and epidermis.

Tentacles: buccal podia extended from the mouth for feeding.

Terminal: occurring at the very posterior end, facing directly posteriorly.

Total allowable catch (TAC): the total number or weight of animals that are legally permitted to be collected or fished in a season or year.

Tranverse: across the body, perpendicular to the main axis of the body.

Trepang: Malaysian name for sea cucumber, also used for the processed product (see also bêche-de-mer).

Trivium: the ventral surface of body in the pentaradiate symmetry, with 3 radii and 2 interradial areas.

Ventral: on the bottom, or under surface, of the animal.

Actinopyga agassizii (Selenka, 1867)

COMMON NAME: Pepino de mar (Costa Rica).

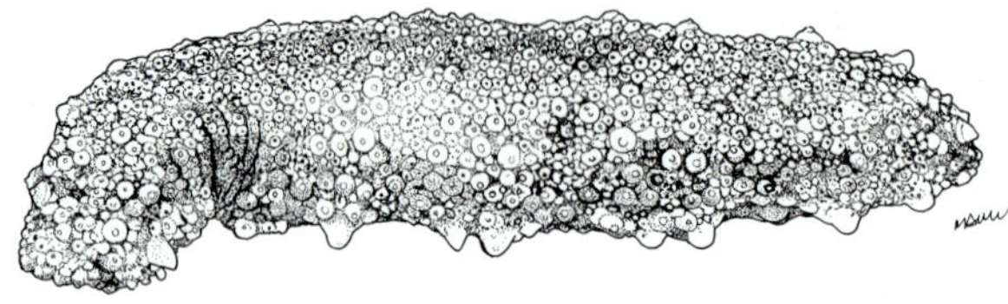

DIAGNOSTIC FEATURES: Body coloration is mottled with brown, yellow and orange patches. Tegument is thick and leathery. The dorsal surface is rounded and covered with numerous wart-like papillae. The ventral surface is flattened, soft and contains broad rows of podia. Ventral mouth with 20 to 30 peltate tentacles. Anus with five calcareous teeth. This species possesses non-adhesive Cuvierian tubules that cannot elongate.

Ossicles: Tentacles with rods of various size, 15–355 µm long, and with the distal ends perforated or not, but always spiny with the sides smooth or also spiny. Dorsal and ventral body wall with the same type of ossicles, which vary from rosettes to simple "dog-biscuits", 30–60 µm long. Ventral podia with huge end-disc. Dorsal podia with X-shaped ossicles.

Processed appearance: Not available.

Size: Maximum length about 35 cm.

HABITAT AND BIOLOGY: Nocturnal species; during the day, it seeks refuge in coral heads, rubble or seagrass beds. *Actinopyga agassizii* can be found from 0 to 54 m deep. It forages on fine detrital sediments in algal turfs, seagrass beds and in rubble or sand-covered areas. The commensal pearl fish, *Carapus bermudensis*, is often found inside the posterior portion of the digestive tract or respiratory tree. In the Bahamas, this species reproduces annually in July and August.

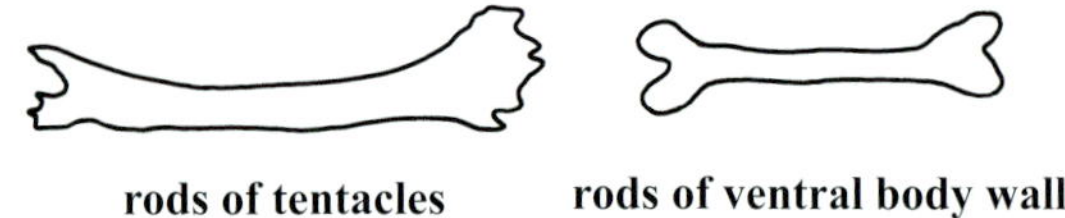

rods of tentacles   rods of ventral body wall

(after Hasbún and Lawrence, 2002)

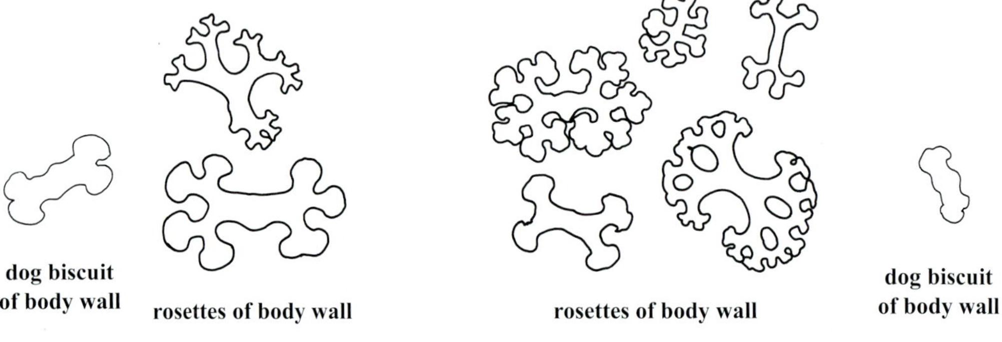

dog biscuit of body wall   rosettes of body wall   rosettes of body wall   dog biscuit of body wall

(source: Hendler *et al.*, 1995)   (source: photo M. Benavides-Serrato)

EXPLOITATION:

Fisheries: Artisanal fishery. In Nicaragua, this species is harvested, without any regulation, with other sea cucumber species. There is commercial exploitation for bêche-de-mer in Panama and Venezuela (Bolivarian Republic of) however, no recent information is available.

Regulations: In Panama, there is a ban on commercial catches of all sea cucumbers (H. Guzman, personal communication), including *A. agassizii*. There is no management of the fishery in Costa Rica.

Human consumption: Consumed as bêche-de-mer.

Main markets and value: Undetermined.

LIVE (photo by: J.J. Alvarado)

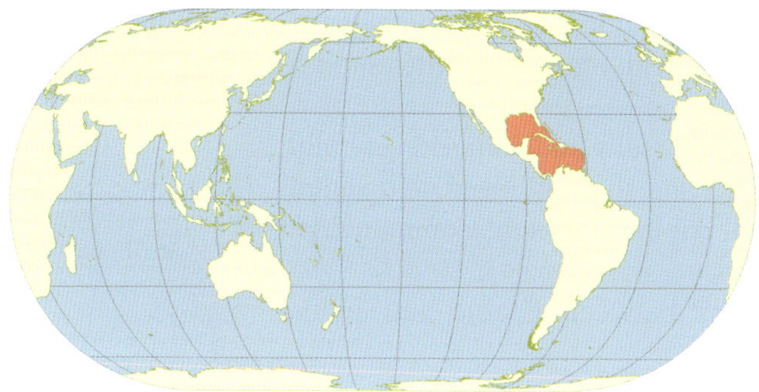

GEOGRAPHICAL DISTRIBUTION:

Caribbean coast of Florida (USA), Cuba, Mexico, Puerto Rico, Dominican Republic, Haiti, Jamaica, Belize, Guatemala, Honduras, Nicaragua, Costa Rica, Panama, Colombia (Atlantic), Venezuela (Bolivarian Republic of), the Bahamas, Barbados and the United States of America.

Actinopyga echinites (Jaeger, 1833)

COMMON NAMES: Deep-water redfish (**FAO**, Papua New Guinea, India, Mauritius, Viet Nam and Madagascar), Brownfish (Réunion), Trokena (Madagascar), Barbara (Mauritius), Pal attai (India), Hải sâm mít (Viet Nam). Also, Hud-hud, Brown beauty, Buli-buli, Khaki, Uwak (Philippines), Goma attaya (Sri Lanka), le Rouge (New Caledonia), Telehea loloto (Tonga), Dri tabua (Fiji).

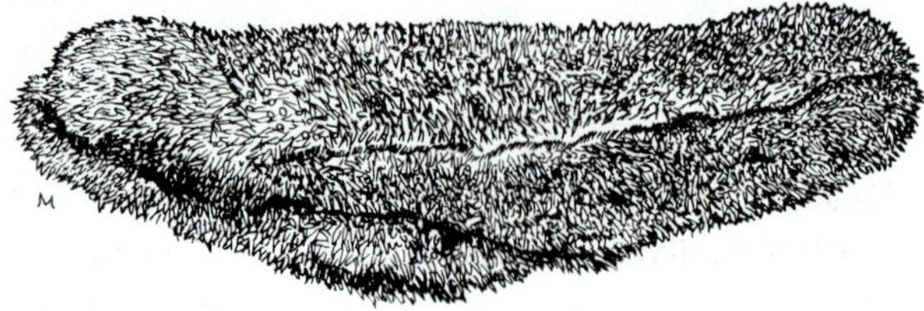

DIAGNOSTIC FEATURES: Colour variable from beige to rusty-brown or dark brown, sometimes with fine dark marks dorsally among papillae. Body moderately elongated. Dorsal surface arched, covered with "pimply" papillae, a couple mm long, sometimes wrinkled and often covered by fine sediments. Ventral surface with numerous long yellow to green podia. Body is flattened ventrally, tapering slightly towards both ends. Mouth is ventral with 20 stout, brown tentacles. Anus surrounded by five small yellow, conical anal teeth. Its small pinkish Cuvierian tubules are generally not extended.

Ossicles: Tentacles with rods, 60–375 μm long, straight or slightly arched, with the ends spiny. Roughly the same types of rods and rosettes are found in the dorsal body wall (20–135 μm long) and ventral body wall (25–80 μm long). Ventral podia with rods and rosettes similar to those of body wall, 20–100 μm long. Dorsal podia with rosettes only.

Processed appearance: Oval shape, arched dorsally and flattened ventrally; with rounded ends. Grey-brown dorsally, rough and slightly ridged, and ventral surface is lighter in colour and granular. Small cut in the mouth. Common dried size 8–15 cm.

Remarks: Some morphological differences between Pacific and Indian Ocean populations.

Size: Maximum length to about 35 cm, commonly to about 20 cm; average fresh weight in Indian Ocean from 200 g (Mauritius) to 300 g (Réunion, Madagascar, India); average weight in New Caledonia about 345 g; average fresh length from 15 (Mauritius) to 20 cm (Réunion, Madagascar, Papua New Guinea, India).

HABITAT AND BIOLOGY: Despite being named "deep-water" redfish, the species actually lives in shallow waters, mostly on flats (reefs and seagrass beds) down to 10 m depth with relatively high densities of up to 1 ind. m^{-2}. This species has separate sexes and is reported to live more than 12 years. Spawning occurs in the dry season, size of maturity is reported at about 12 cm, or a weight between 45 g and 90 g.

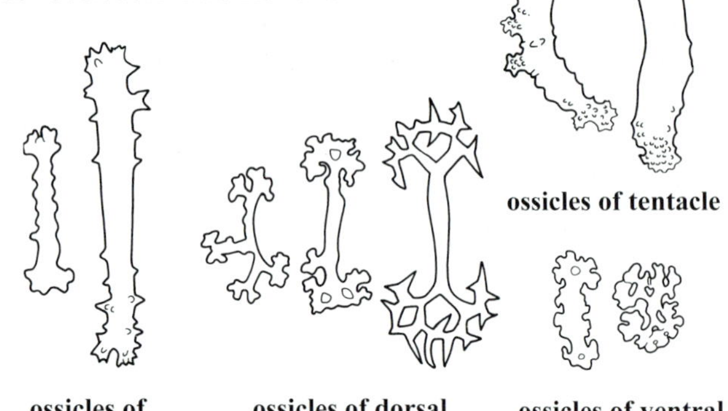

ossicles of tentacle

ossicles of ventral podia

ossicles of dorsal body wall

ossicles of ventral body wall

(after Féral and Cherbonnier, 1986)

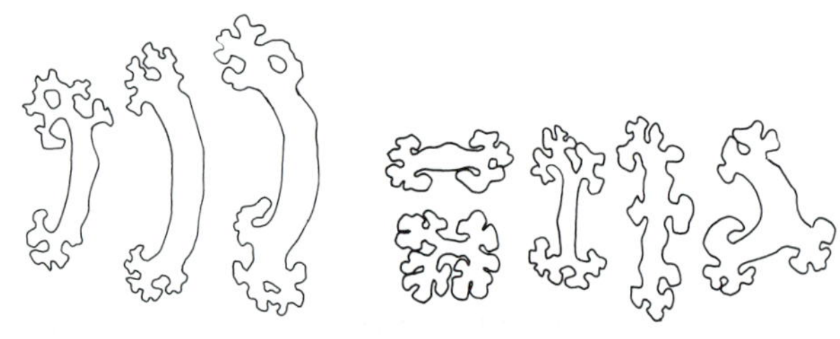

rods of ventral body wall

rosettes of ventral body wall

(after Panning, 1944)

EXPLOITATION:

Fisheries: Fished artisanally throughout its distribution. Collected by hand at low tide or by skin diving (e.g. Zanzibar [Tanzania] and Madagascar), through the use of lead-bombs (Papua New Guinea), using SCUBA diving (Mauritius) and hookah gear (Viet Nam). It is fished heavily in Sri Lanka and other Asian countries and throughout much of the Central Western Pacific.

Regulations: On the Great Barrier Reef (Australia), the minimum size limit (wet) is 15 cm and this species is subject to rotational zone closures and a total allowable catch (TAC) limit. Before a moratorium on the fishery in Papua New Guinea, there was a minimum size limit (25 cm live; 15 cm dry) and other regulatory measures. Fishing for this species is banned by moratoria in numerous countries.

Human consumption: Mostly, the reconstituted body wall (bêche-de-mer) is consumed by Asians.

Main markets and value: Singapore, Hong Kong China SAR, China, Taiwan Province of China. In New Caledonia, this species is exported for USD20–30 kg^{-1} dried and fishers may receive USD2 kg^{-1} wet weight. It is traded at USD28–54 kg^{-1} dried in the Philippines. Wholesale prices in Guangzhou were up to USD63 kg^{-1}.

LIVE (Pacific Ocean variety) (photo by: S.W. Purcell)

LIVE (Indian Ocean variety) (photo by: P. Bourjon)

PROCESSED (Pacific Ocean variety) (photo by: S.W. Purcell)

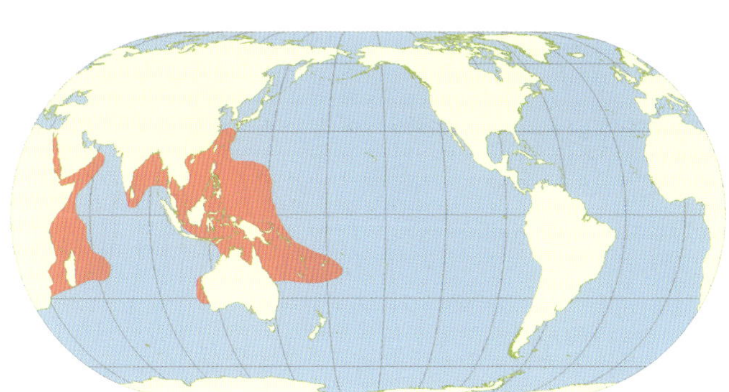

GEOGRAPHICAL DISTRIBUTION:

Found throughout the western central Pacific, Asia, Africa and Indian Ocean region. Common in the Indo-Pacific, islands of western Indian Ocean, Mascarene Islands, East Africa, Madagascar, southeast Arabia, Sri Lanka, Bay of Bengal, East Indies, north Australia, the Philippines, China and southern Japan, South Sea Islands.

OTHER USEFUL INFORMATION: Several authors report this species as overexploited in many areas due to its desirability in the international trade.

COMMON NAMES: Stonefish, Pal attai (India), Mbura (Zanzibar. Tanzania), Buliq-buliq, Monang, Munang (Philippines), Le caillou (New Caledonia), Telehea maka (Tonga), Dri vatu (Fiji).

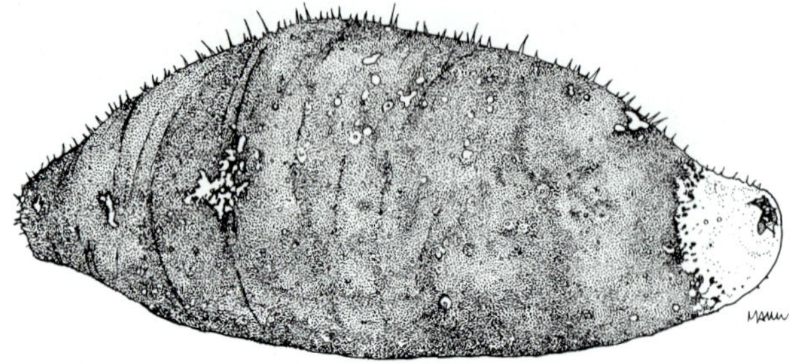

DIAGNOSTIC FEATURES: Body is nearly uniformly beige to chocolate brown with some lighter spots and sometimes fine dark blotches. Around the anus, it is usually characteristically white. The ventral surface of smaller individuals is usually beige. Body is stout and tapered at both ends. Dorsal surface is quite arched, while the ventral surface is flattened. Few papillae scattered over dorsal area. Ventral mouth with greenish-brown or brown tentacles. Anus is terminal with five strong, yellowish teeth. Cuvierian tubules absent.

Ossicles: Tentacles with massive rods, 45–450 µm long, straight or slightly arched, and spiny at the extremities. Dorsal body wall with small rosettes, 25–35 µm long, or larger X-shaped rosettes, 50 µm long. Ventral body wall with even smaller rosettes, 20–25 µm long. Ventral podia with tiny rosettes, 10–25 µm long. Dorsal podia with rosettes similar to those of the body wall and rods, 65–90 µm long.

Processed appearance: Roughly oval shape; arched dorsally, mildly flattened ventrally. Brown-black; dosally with shallow grooves, while ventral surface is smooth. Common dried size between 10 and 12 cm.

Remarks: Body is firm like a stone when handled, giving rise to its common name.

Size: Maximum length about 24 cm; average fresh weight 400 g; average fresh length about 20 cm.

HABITAT AND BIOLOGY: This species lives in coral and coral rocks and reef ledges, between 0.5 and 7 m. It prefers hard substrates (i.e. coral reefs) that are sheltered. It is a predominantly nocturnal species. During the day, it seeks shelter under large stones and reef crevices. In Papua New Guinea, it can be found in waters up to 20 m deep.

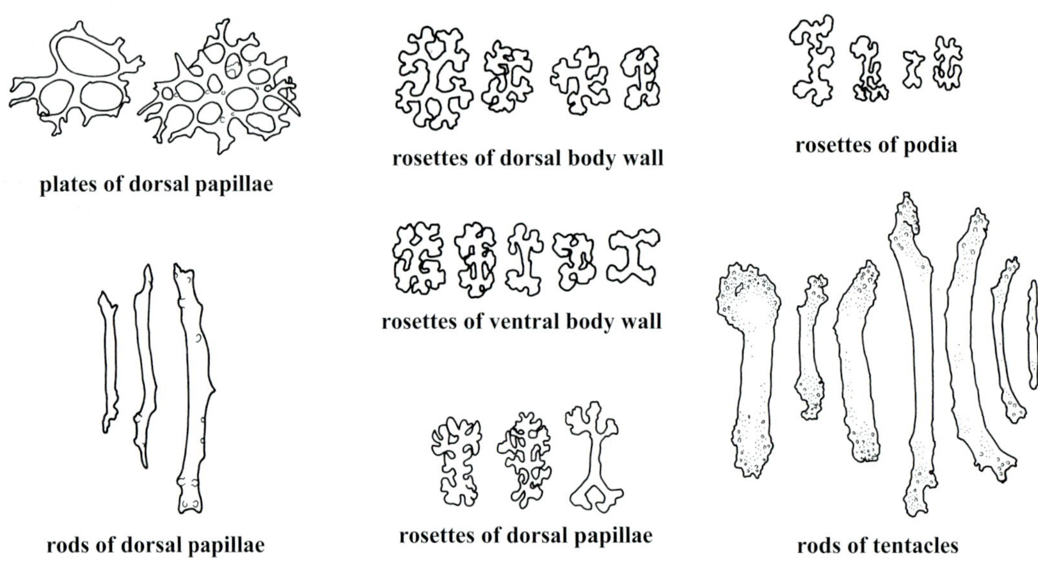

plates of dorsal papillae

rosettes of dorsal body wall

rosettes of podia

rosettes of ventral body wall

rods of dorsal papillae

rosettes of dorsal papillae

rods of tentacles

(after Massin, 1999)

EXPLOITATION:

Fisheries: Harvested in artisanal fisheries throughout its whole distribution range where it is hand collected, using lead-bombs and free-diving. In some fisheries, it is collected with torches at night by SCUBA diving or free-diving (e.g. Philippines and Viet Nam) but this practice is often banned in other fisheries.

Regulations: Before a fishery moratorium in Papua New Guinea, fishing for this species was regulated by minimum landing size limits (15 cm live; 10 cm dry) and other regulations.

Human consumption: Mostly, the reconstituted body wall (bêche-de-mer) is consumed by Asians. Its intestine and/or gonads may be consumed as part of traditional diets.

Main markets and value: Marketed in Singapore, India, Papua New Guinea and China; in the latter it is considered of medium low commercial importance only. It is traded at USD20–6 kg^{-1} dried in the Philippines. In Papua New Guinea it was previously sold by fishers at about USD7 kg^{-1} dried. In Fiji, fishers receive USD5–8 per piece fresh. Prices in Guangzhou wholesale markets ranged from USD79 to 108 kg^{-1}.

LIVE (photo by: S.W. Purcell)

PROCESSED (photo by: S.W. Purcell)

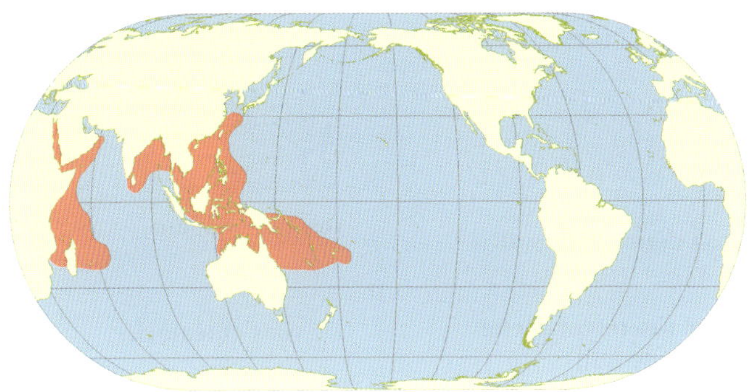

GEOGRAPHICAL DISTRIBUTION:

The Mascarene Islands, East Africa to the Red Sea and Oman, Madagascar, Sri Lanka, Bay of Bengal, East Indies, north Australia, the Philippines, China and southern Japan, South Pacific Islands. In India, it is found only in Andamans and Lakshadweep regions.

Actinopyga mauritiana (Quoy and Gaimard, 1833)

COMMON NAMES: Surf redfish (**FAO**), Holothurie brune des brisants (**FAO**), Pal attai (India), Đồn đột dừa (Viet Nam), Fotsisetsake (Madagascar), Kajno (Egypt), Mbura (Zanzibar, Tanzania), Yellow surfish (Seychelles), Bakungan, Monang (Philippines), La mauritiana (New Caledonia), Terasea (Fiji), Tewaeura (Kiribati), Telehea kula (Tonga).

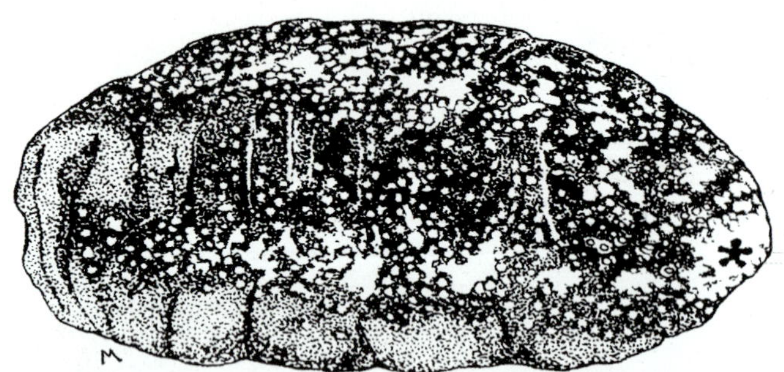

DIAGNOSTIC FEATURES: Body coloration is variable from brown to reddish and often with numerous white spots dorsally on individuals in the Pacific. In the Western Indian Ocean, the coloration is marbled greenish to brownish with white patches dorsally. Body is stout, arched dorsally and flattened ventrally, and mildly tapering at the ends. Dorsal surface may be wrinkled and pitted and seldom covered by sediment. Slippery to the touch. Papillae on dorsal surface are slender and short. Ventral podia are densely packed and greenish to light brown. Mouth ventral with 25 short and stout, brown tentacles. Anus surrounded by five teeth, often whitish. Cuvierian tubules absent.

Ossicles: Tentacles with large very rugose rods, 165–210 µm long. Dorsal body wall with spiny rods, 55–90 µm long and simple rosettes, 20–45 µm long. Ventral body wall with small grains, elongated grains and rods that can be spiny and are 20–80 µm long. Ventral podia devoid of ossicles, apart from the end-plate. Dorsal papillae with rods with slightly spiny or ragged sides, 100 µm long on average, as well as large rosettes, 50–60 µm long.

Processed appearance: Roughly an elongated oval shape; arched dorsally and moderately flattened ventrally. Dorsal surface with grooves, blackish brown and showing the former white spots and blotches. Ventral surface is granular and characteristically cream to light reddish brown in colour. A cut on the ventral surface. Common dried size 8–15 cm.

Remarks: This species adheres tightly to wave-exposed surfaces with its numerous ventral podia.

Size: Maximum length about 35 cm, commonly to 20 cm; average fresh weight from 300 to 700 g; average fresh length from 20 to 40 cm. In New Caledonia, average live weight about 670 g and average live length about 20 cm.

HABITAT AND BIOLOGY: Prefers outer reef flats and fringing reefs, in very shallow waters, near low water mark where surf breaks, generally in 1–3 m water depth. Reportedly found occasionally in seagrass beds, attached to coral stones. Can be active in both daytime and night. It feeds on detritus lying on hard reef substrates, and its movement can be affected by tides.

Size at maturity is reported at 23 cm, reached between 125 and 350 g.

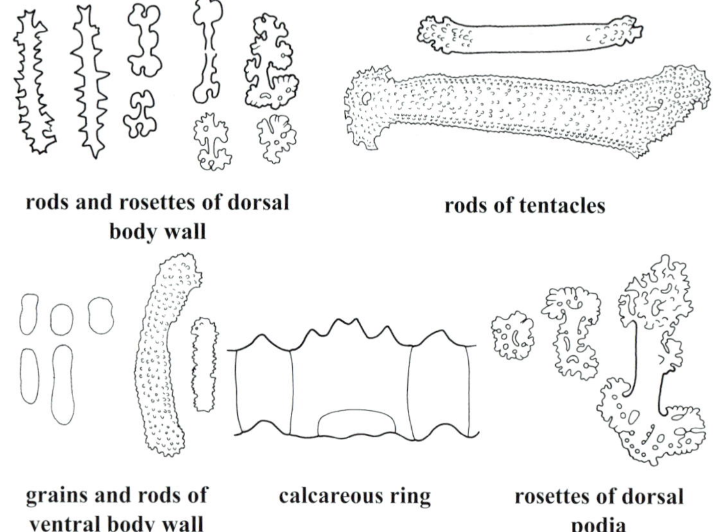

rods and rosettes of dorsal body wall

rods of tentacles

grains and rods of ventral body wall

calcareous ring

rosettes of dorsal podia

(after Féral and Cherbonnier, 1986)

EXPLOITATION:

Fisheries: Fished throughout the Indo-Pacific. It is favoured for subsistence consumption in Palau, Nauru, Wallis and Futuna Islands, Samoa, Cook Islands, French Polynesia. Harvested by hand and by free-diving in at least 22 countries and island States in the western central Pacific. In Egypt, depletion of populations has prompted the study and possible development of aquaculture-based restocking. It is of commercial importance in southern China, Japan, Malaysia, Indonesia and the Philippines. It was an important species in Northern Mariana Islands before the fishery was closed. In Seychelles, it is one of the most important commercial species, where it is considered to be overexploited. It is heavily exploited in Tonga.

LIVE (Pacific Ocean variety) (photo by: S.W. Purcell)

PROCESSED (photo by: S.W. Purcell)

Regulations: Before a fishery moratorium in Papua New Guinea, fishing was regulated by minimum landing size limits (20 cm live; 8 cm dry) and other regulations. In New Caledonia, fishing for this species is prohibited at night and there is a minimum legal size limit (25 cm, fresh animals; 12 cm, dried). On the Great Barrier Reef (Australia), a minimum size limit of 25 cm (wet/fresh) applies to this species, which is also subject to a rotational closure strategy, permits and a quota. In Torres Strait (Australia), the *Actinopyga mauritiana* fishery was closed in 2003 due to overexploitation.

Human consumption: Mostly, the reconstituted body wall (bêche-de-mer) is consumed by Asians. Its intestines and/or gonads may be consumed as part of traditional diets or in times of hardship.

Main market and value: China, Singapore, Hong Kong China SAR, Taiwan Province of China and Ho Chi Minh City Chinese markets for further export. It is traded at USD13–47 kg^{-1} dried in the Philippines. In Papua New Guinea, it was previously sold at USD10–15 kg^{-1} dried. In New Caledonia, it is exported for about USD30 kg^{-1} dried and fishers may receive USD2 kg^{-1} wet weight. In Fiji, fishers receive USD2–4 per piece fresh. Prices in Guangzhou wholesale markets ranged from USD57–79 kg^{-1}. Retail prices in Hong Kong China SAR were up to USD145 kg^{-1} dried.

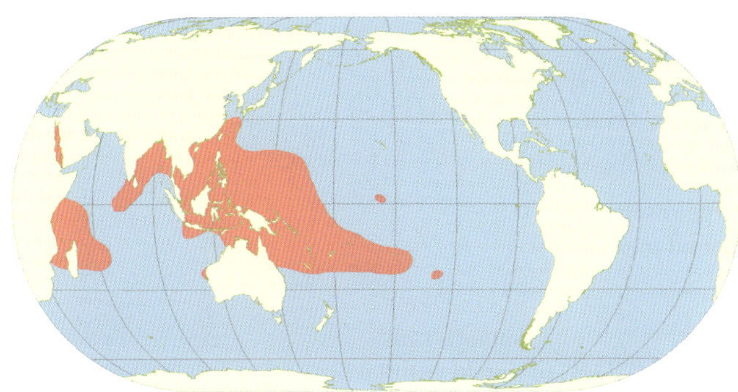

GEOGRAPHICAL DISTRIBUTION:

Islands of western Indian Ocean, Mascarene Islands, East Africa, Madagascar, Red Sea, Maldives, Sri Lanka, Bay of Bengal, East Indies, north Australia, the Philippines, China and southern Japan, South Pacific Islands (see SPC PROCFish/C surveys) as far east as Pitcairn Islands. In India, it is distributed in the Gulf of Mannar, the Andamans and Lakshadweep.

Actinopyga miliaris (Quoy and Gaimard, 1833)

COMMON NAMES: Blackfish, Hairy blackfish (**FAO**), Pal attai (India), Khaki (Philippines), Kalu attaya (Sri Lanka), La boule noire (New Caledonia), Loli fulufulu (Tonga), Dri Loli (Fiji), Kijini (Zanzibar, Tanzania).

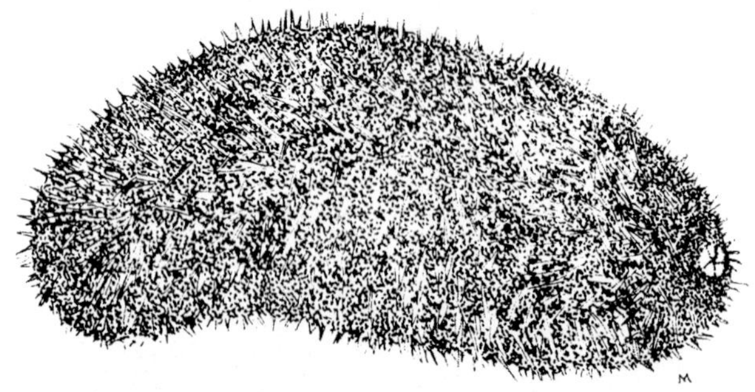

DIAGNOSTIC FEATURES: This species is brown to blackish dorsally, and lighter brown ventrally. Body stout, cylindrical, slightly arched dorsally and somewhat flattened ventrally. It has long slender podia dorsally, making it look 'hairy'. The dorsal surface is generally covered by mucus and may have fine sediment. Mouth ventral surrounded by 20 stout, brown to black, tentacles. Anus surrounded by 5 strong, conical or bicusped anal teeth that are yellow to orange. Cuvierian tubules not present.

Ossicles: Tentacles with large quite spiny rods, 30–300 µm long, more spiny distally. Dorsal and ventral body wall with similar rosettes of the same length ± 25 µm. Ventral podia with few rosettes, which are similar to those of the body wall. Dorsal papillae with rosettes of the same size as those in the body wall, and some larger ones, 45–80 µm long. Dorsal papillae also have two types of rods: spiny, 55–230 µm long, ones, often with numerous lateral spiny extensions, and smooth 100–150 µm long, sometime rosette-like ones.

Processed appearance: Roughly oval shape with a round cross-section. Entire body surface smooth and black. A small cut may be made across the mouth or across the ventral surface. Common size 10–12 cm.

Remarks: The lighter brown ventral surface of *Actinopyga miliaris* and its more simple anal teeth distinguish it from *A. spinea* and *A. palauensis*. This species usually contracts to a sphere or stout rugby-ball shape when handled, unlike the other two species with which it may be confused.

Size: Maximum length about 35 cm, commonly to about 25 cm; average fresh weight 400 g; average fresh length 25 cm.

HABITAT AND BIOLOGY: It is distributed commonly between 0 and 10 m deep, on sandy beds and intertidal areas. In the western central Pacific region, found mostly on reef flats of fringing reefs and lagoon-islet reefs between 0 and 12 m depth. In the African and Indian Ocean region, it prefers reef flats and seagrass beds over coral substrate up to 20 m and it does not bury. In China, it reportedly prefers areas affected by a strong wave action. In New Caledonia, it does bury in some localities and

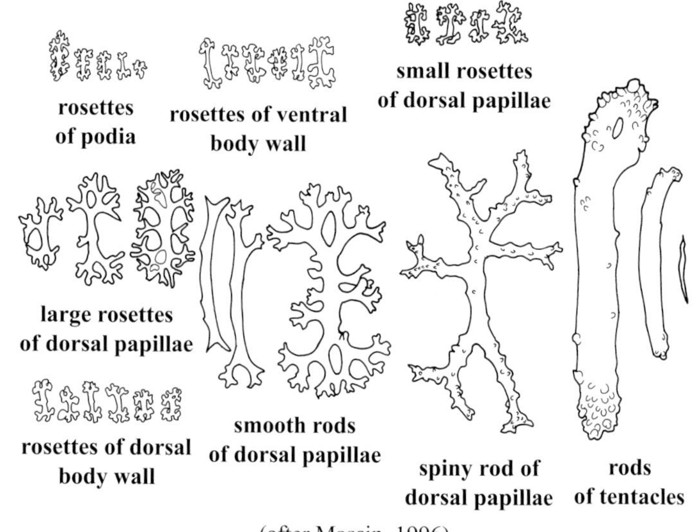

(after Massin, 1996)

reproduces twice a year, with one spawning event in May and a second in November and December.

EXPLOITATION:

Fisheries: In the western central Pacific region, this species is harvested in more than a dozen countries. In Fiji, *A. miliaris* was among the most important commercial species before 1988, accounting for about 95% of all exports. In Asia, this species is a heavily fished species in certain countries of its distribution range such as China, Indonesia and the Philippines. It is actively fished in Kenya, representing about 17% of total catches. In Tanzania, there was intense fishing of this species in the early 1980s but yields declined subsequently.

Regulations: Before a fishery moratorium in Papua New Guinea, fishing for this species was regulated by minimum landing size limits (20 cm live; 8 cm dry) and other regulations. Since late 2007, there has been a fishery moratorium in Yap for *A. miliaris*. In New Caledonia, the permissible size is 25 cm live and 12 cm dry. The minimum legal length on the Great Barrier Reef is 20 cm, whereas in Torres Strait (Australia) it is 22 cm.

Human consumption: Mostly, the reconstituted body wall (bêche-de-mer) is consumed by Asians. Its intestines and/or gonads may be consumed as part of traditional diets or in times of hardship, e.g. in Palau.

Main market and value: Singapore, Asia. It is sold at about USD15 kg^{-1} dried. In New Caledonia, fishers may receive USD2.4 kg^{-1} wet weight. In Fiji, fishers receive USD1–5 per piece fresh. Prices in Guangzhou wholesale markets ranged from USD63 to 92 kg^{-1} dried.

LIVE (photo by: S.W. Purcell)

PROCESSED (photo by: S.W. Purcell)

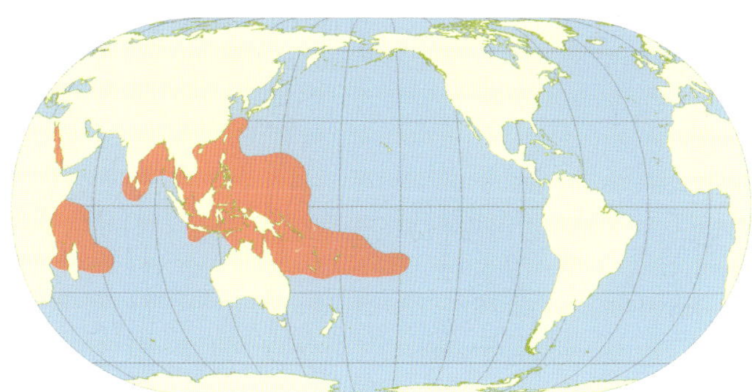

GEOGRAPHICAL DISTRIBUTION:

Islands of western Indian Ocean, Mascarene Islands, East Africa, Madagascar, Red Sea, Sri Lanka, Bay of Bengal, East Indies, north Australia, the Philippines, China and southern Japan, South Pacific Islands east to French Polynesia. In India, it is known from the Gulf of Mannar, Palk Bay, the Andamans and Lakshadweep.

COMMON NAMES: Deepwater blackfish, Panning's blackfish (**FAO**), Le noir long (New Caledonia).

DIAGNOSTIC FEATURES: Glossy brownish-black uniformly over body. Dorsal surface bumpy and the upper part of the dorsal surface is usually covered by some coarse sand (but body not completely covered in sand, c.f. *Holothuria whitmaei*). Body subcylindrical, flattened slightly ventrally. Buccal cavity (mouth) often projected (trunk-like). Tentacles brown. Anus terminal with five prominent multidentate teeth that may be flattened or club-shaped with small knobs. Dorsal papillae small and sparse; ventral papillae relatively numerous.

Ossicles: Tentacles with rods of various size and rugosity, the smallest ones, 80 µm long, nearly smooth while the largest ones are distantly spined and up to 700 µm long. Dorsal and ventral body wall with the same type of ossicles: branching, quite asymmetrical, non-perforated rods, 25–75 µm long. Ventral podia with simpler ragged rods, generally somewhat shorter than those of the body wall. Dorsal podia with rods that are less ragged and more curved, up to 85 µm long.

Processed appearance: Dark brown to black in colour. Cylindrical in shape with relatively blunt ends, like a salami, and not particularly larger towards the middle of the body (c.f. *Actinopyga miliaris*). Somewhat bumpy (textured) and finely wrinkled on the dorsal surface. Common dried size 15–20 cm.

Remarks: Often grouped with *A. spinea* by fishers and processors. Probably often misidentified as *A. miliaris* or *A. spinea* in visual surveys. This species has been combined with counts of *A. miliaris* in underwater surveys from the Pacific, so its abundance and distribution has been underestimated.

Size: Maximum length about 40 cm; average fresh weight: 1 600 g; average fresh length: 27 cm. In New Caledonia, average live weight about 1 450 g and average live length about 25 cm.

HABITAT AND BIOLOGY:

Commonly found on deeper hard reef surfaces and coarse sand with coral rubble. More common on reef slopes of outer reef passes, with clear water and fore-reef pavement. It can also be found on semi-sheltered bay reefs with boulders and coral rubble. Seldom on lagoon reefs or inshore reefs. Often on open surfaces and probably buries little. Occurs in 4 to 25 m depth and is non-cryptic.

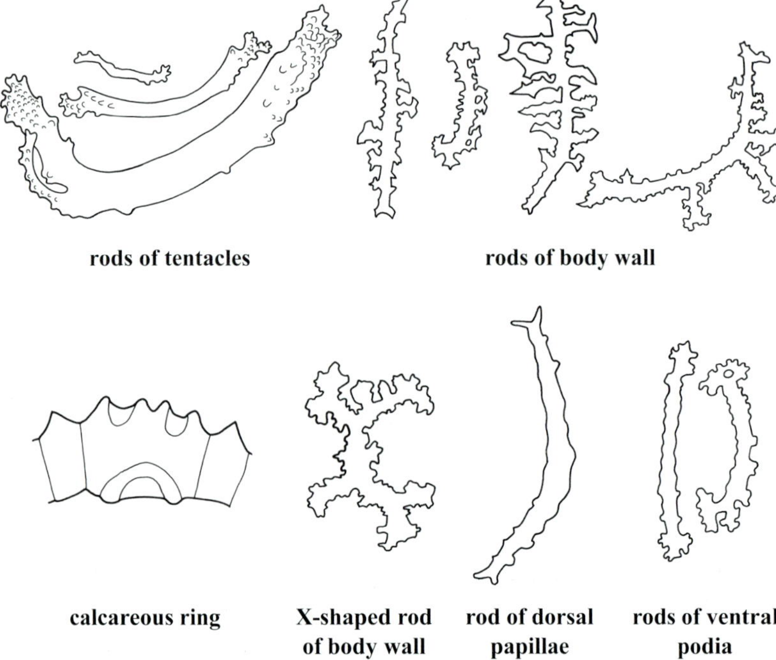

rods of tentacles

rods of body wall

calcareous ring

X-shaped rod of body wall

rod of dorsal papillae

rods of ventral podia

(after Cherbonnier and Féral, 1984)

EXPLOITATION:

Fisheries: Harvested by skin divers for artisanal fisheries. Also targeted in semi-industrial and industrial fisheries. This species is fished in Australia, Palau, Micronesia (Federated States of), Kiribati, Fiji, Tonga, Niue and New Caledonia. In the last, it is among the dominant species in the catches. In the State of Yap, part of Micronesia (Federated States of), there is a potential to further develop the fishery for *A. palauensis*.

Regulations: On the Great Barrier Reef (Australia), fishing of this species is regulated by a global TAC, a rotational harvest strategy and fishing permits. In New Caledonia, fishing permits are required and SCUBA is prohibited for fishing this species.

Human consumption: Mostly, the reconstituted body wall (bêche-de-mer) is consumed by Asians.

LIVE (photo by: S.W. Purcell)

PROCESSED (photo by: S.W. Purcell)

Main market and value: In New Caledonia, it is exported for about USD35 kg^{-1} dried and fishers may receive USD2.4 kg^{-1} wet weight. In Fiji, fishers receive USD2–16 per piece fresh. Prices in Guangzhou wholesale markets ranged from USD95 to 116 kg^{-1} dried.

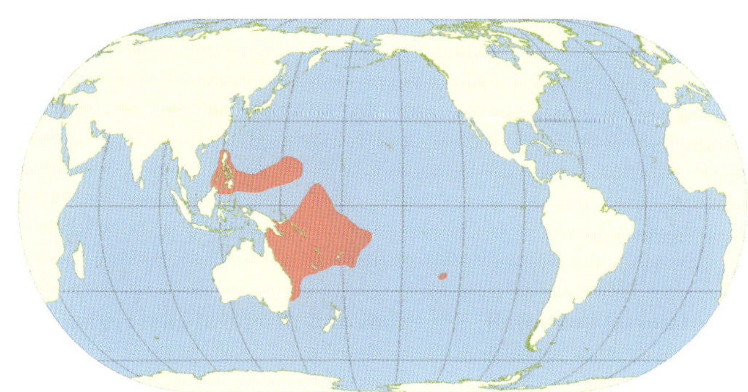

GEOGRAPHICAL DISTRIBUTION:

East Australia, as far south as Solitary Islands and Lord Howe Island (S. Purcell personal observation), and the western Pacific Islands to the Pitcairn Islands group; reported in various countries of Melanesia and Micronesia. This species has probably been recorded often as *A. miliaris* (e.g. SPC PROCFish/C surveys) or other species, leading to an under-represented distribution.

OTHER USEFUL INFORMATION: This species can be distinguished from *A. miliaris* by having relatively sparse, small, dorsal podia and it does not retract its body into a sphere when handled. It has multidentate anal teeth (whereas more-or-less simple and conical in *A. miliaris*) that may have terminal serrations on an axis. It can be distinguished from *A. spinea* in being black-brown (appearing black at depth), the anus is more terminal, and the dorsal surface is noticeably textured.

COMMON NAMES: Burying blackfish, New Caledonia blackfish (**FAO**), Le noir long (New Caledonia).

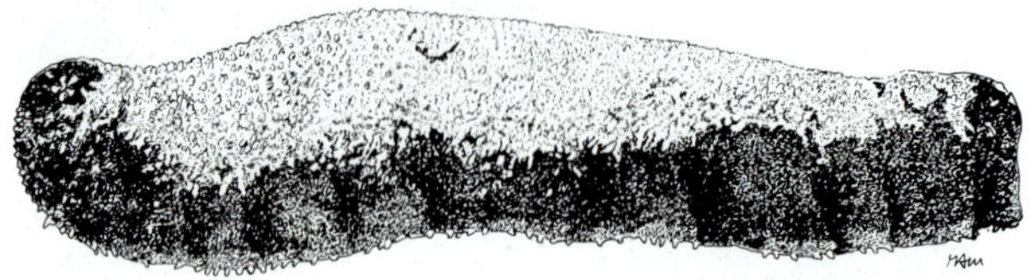

DIAGNOSTIC FEATURES: Coloration is rusty brown to dark brown, or brownish-black, and uniform over entire body. Sometimes covered by a dusting of fine sand. Body subcylindrical and flattened slightly ventrally. It may be quite elongated compared with other species within this genus. The 20 tentacles are dark brown. Moderately long, but thin, dorsal papillae that are less numerous than the short ventral podia. Anus subdorsal with five prominent, yellowish, nodular teeth. Cuvierian tubules absent.

Ossicles: Only the peristome, the anal region and the tentacles are rich in ossicles. Tentacles with spiny curved rods of various size, 250–500 µm long. Dorsal body wall with few forked spiny rods, about 110 µm long, and spiny plates of various size, 80–130 µm long. Ventral body wall devoid of ossicles. Ventral podia with few short, approximately 120 µm long, bifurcating rods. Dorsal podia with similar rods, but twice as long.

Processed appearance: Elongated and cylindrical. Colour is dark brown and texture is relatively smooth.

Remarks: Often combined with *Actinopyga miliaris* by fishers and processors and probably often misidentified as *A. miliaris* in visual resource surveys. Incorrectly combined with counts of *A. miliaris* in underwater surveys from the Pacific, so its abundance and distribution has been underestimated.

Size: Maximum length about 38 cm, commonly to about 27 cm; average fresh weight: 700 g; average fresh length: 25 cm. In New Caledonia, average live weight about 1 040 g and average live length about 25 cm.

HABITAT AND BIOLOGY:
Commonly on muddy-sand habitats, where it buries. It can occur in 1 to 25 m depth in other sandy habitats alongside *A. miliaris*, including sandy reef flats and protected fine sand lagoons and bays.

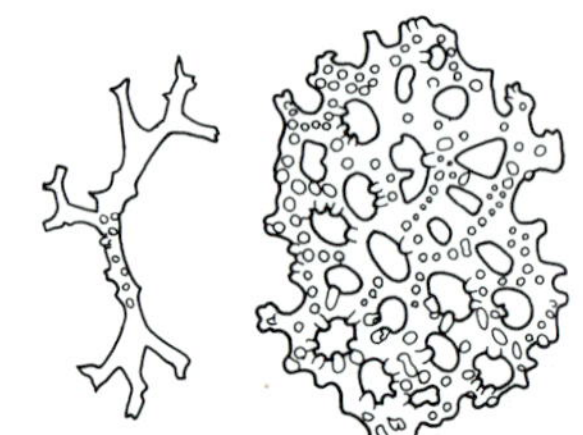

rod and plate of dorsal body wall

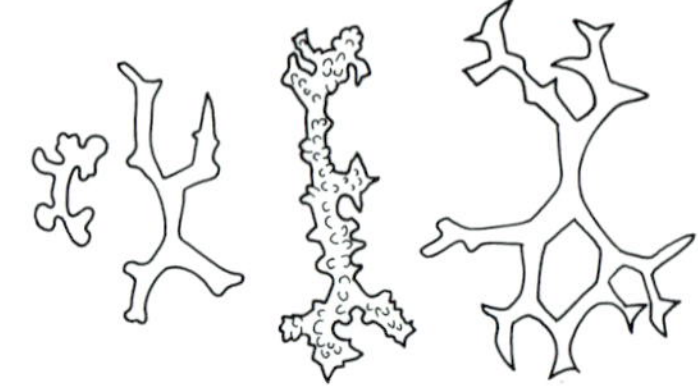

rods and rosette of anal region

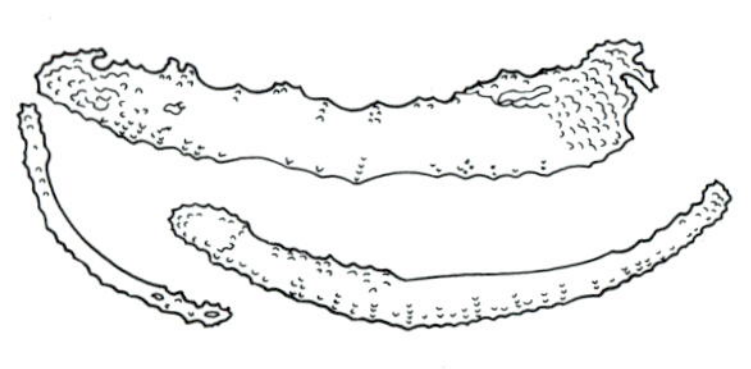

rods of tentacles

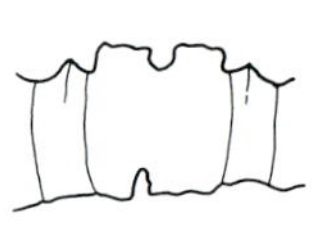

calcareous ring

rod of ventral podia

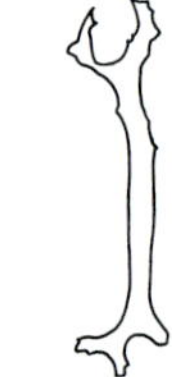

rod of dorsal papillae

(after Cherbonnier, 1980)

EXPLOITATION:

Fisheries: Artisanal and semi-industrial. This species is fished in the Great Barrier Reef (Australia) and New Caledonia. In the former, it is currently the dominant species in catches. Collection in New Caledonia is by free-diving and gleaning on reefs at low tide, whereas on the Great Barrier Reef, it is collected by divers using hookah.

Regulations: In New Caledonia, it is managed with no-take marine reserves, a prohibition on the use of SCUBA or hookah, and fishing permits. In Australia, there is a minimum size limit, a TAC and catches are carefully monitored from specified fishing plots by a small number of licensed fishers. Corrently, from the Great Barrier Reef, more than 150 tonnes of this species is harvested annually and fishers collect >30 kg of animals h^{-1}. Since 2006, *A. spinea* has been distinguished and managed by Queensland authorities separately from *A. miliaris*. It is excluded from the rotational fishing closures used for other species on the Great Barrier Reef.

Human consumption: Mostly, the reconstituted body wall (bêche-de-mer) is consumed by Asians. Sold also as cooked, frozen, product.

Main market and value: It is sold at USD2–4 per unit fresh and USD20–30 kg^{-1} dried. In New Caledonia, fishers may receive USD2.4 kg^{-1} wet weight. Prices in Guangzhou wholesale markets ranged from USD63 to 95 kg^{-1} dried.

LIVE (photo by: S.W. Purcell)

PROCESSED (photo by: S.W. Purcell)

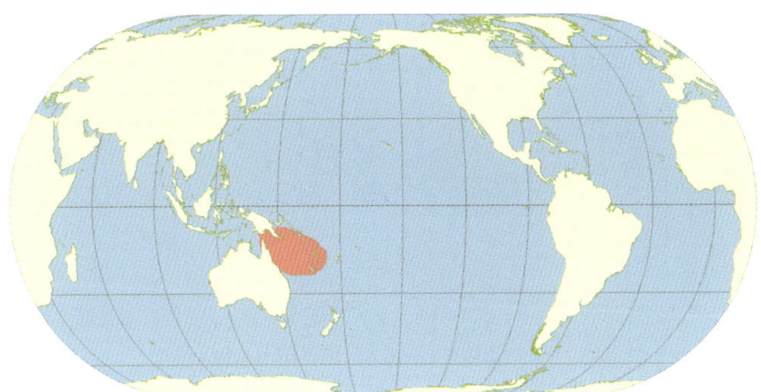

GEOGRAPHICAL DISTRIBUTION:

New Caledonia and Australia. Possibly also occurs in other Melanesian countries but previously misidentified. This species has probably been recorded often as *A. miliaris* (e.g. as in SPC PROCFish/C surveys) or other species, probably leading to an under-represented distribution.

OTHER USEFUL INFORMATION: Ossicles are very sparse in the dermal tissue of this holothurian, a large surface area of dermis is needed for ossicle analysis. Somewhat difficult to distinguish from *A. miliaris*, but key identifying features are that it is often more elongated, anus is subdorsal, the ventral surface is not distinctly lighter than the dorsal surface, and it has only moderately long papillae.

COMMON NAMES: Spiky deep-water redfish.

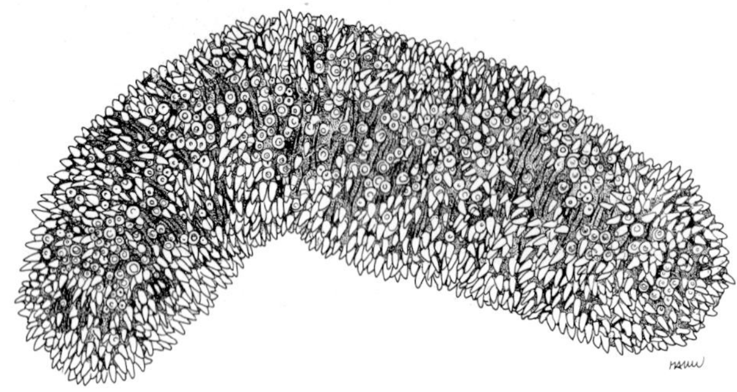

DIAGNOSTIC FEATURES: This species can attain a relatively large size. Arched dorsally and mildly flattened ventrally. Body coloration is orange to pink to flame-red with characteristic blueish to greyish conical papillae over entire dorsal and lateral surfaces of body. The body wall is relatively thick (8–10 mm). Podia are relatively sparse on the ventral surface. The mouth is ventral with 20 brown tentacles. The anus is subdorsal.

Ossicles: Tentacle stalks with very massive rods that are spiny at the extremities and with few perforations, up to 300 µm long. Tentacle tips with more slender and smaller rods, 50–75 µm long. Dorsal body wall with spiky plates, 40–60 µm long, and spiny rods, 50–55 µm long. Ventral body wall with small rods with rounded or spiky ends which may branch to the extent that they become X-shaped; the size of these different ossicles varies from 40–100 µm. Ventral and dorsal podia with small rods as in the ventral body wall as well as rare irregularly shaped perforated rods.

Processed appearance: Relatively elongate and rounded at the ends. Dark coffee-brown in colour. Similar processed shape to *Actinopyga palauensis*, however, body covered with light brown conical papillae. Anal teeth should be evident, which may help to distinguish it from some processed stichopodid species. A small cut across mouth or along middle part of ventral surface.

Remarks: This species is fished in parts of the Central Western Pacific. Genetic samples indicate that the fished species is not *A. flammea* as described by Cherbonnier 1979, but rather a closely related species, which is here tentatively listed as *A.* sp. affinity *flammea*. Populations appear quite sparse and may be at risk of extinction.

Size: Appears to attain at least 2 kg in weight and 45 cm long. Average length is probably 25–30 cm.

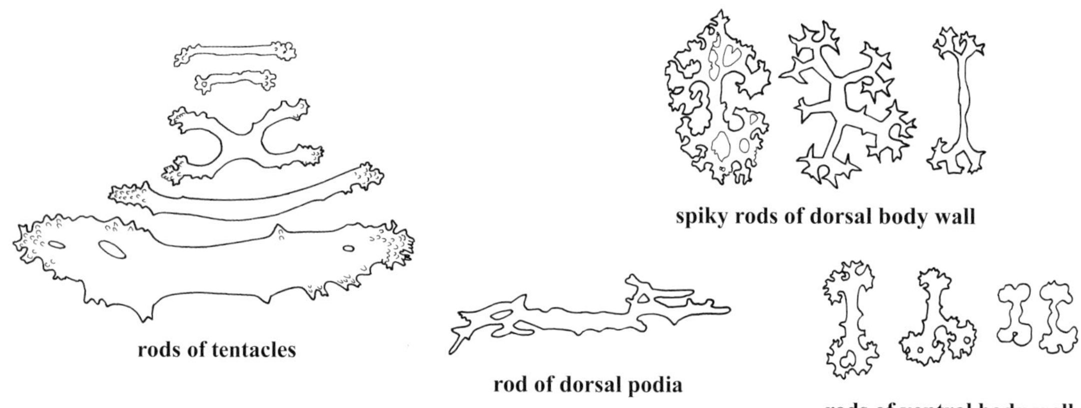

spiky rods of dorsal body wall

rods of tentacles

rod of dorsal podia

rods of ventral body wall

(after Cherbonnier, 1979)

HABITAT AND BIOLOGY: It occurs in deep reef habitats. Anecdotal records suggest this species prefers outer reef slopes and deeper oceanic lagoon habitats. The type specimen was collected in waters at 40–45 m depth on dead coral pavement with brown and crustose-coralline algae. Reported to occur down to 60 m depth.

EXPLOITATION:

Fisheries: It is fished commonly but in low numbers in at least Fiji and Tonga. Very seldom fished in New Caledonia, from where the type specimen was collected for the original description.

Regulations: No specific regulations, such as a species-specific size limit, exist for this species.

Human consumption: This species does not appear to be used in subsistence fisheries. The predominant use is as reconstituted bêche-de-mer.

Main markets and value: Bought by processors in Fiji for about USD5–15 per piece fresh.

LIVE (photo by: G. Paulay)

PROCESSED (photo by: S.W. Purcell)

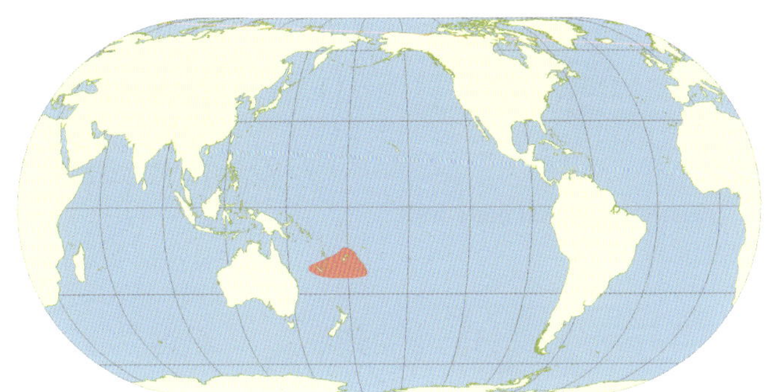

GEOGRAPHICAL DISTRIBUTION: Known at least from New Caledonia, Fiji and Tonga, but does not appear to reach as far north as Kiribati. May occur in the Coral Triangle region.

Bohadschia argus Jaeger, 1833

COMMON NAMES: Leopard fish, Holothurie léopard (**FAO**), Tiger fish, Ñoät da traên, Sam vang (Viet Nam), Nool attai (India), Leopard or Matang itik (Philippines), Tetaika (Kiribati), Mata mata (Tonga), Vulu ika (Fiji).

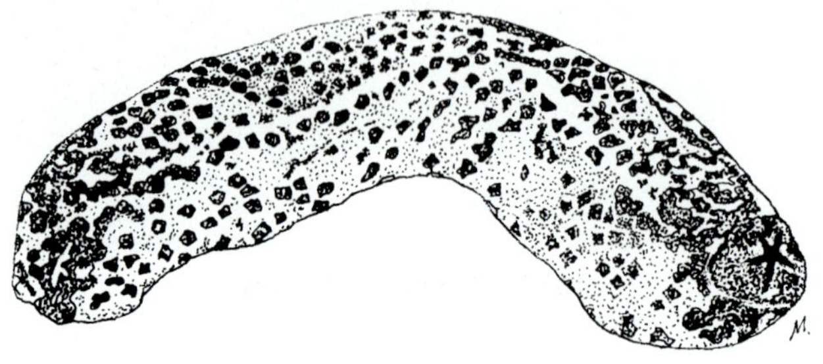

DIAGNOSTIC FEATURES: Variable colour, ranging from brown to beige or grey or violet. Dorsal surface covered with numerous large (about 1 cm) polygonal or round spots, brownish or whitish in colour with a thin black border and sometimes haloed by white. A papilla is at the centre of each spot. Brown to beige ventrally. This species is a moderately large holothurian. Body is cylindrical; arched dorsally and flattened ventrally. Mouth is ventral and surrounded by 20 short dark brown tentacles. Prominent subdorsal anus, no anal teeth. Readily ejects large, white Cuvierian tubules when disturbed.

Ossicles: Tentacles with spiny rods, 80–300 µm long. Dorsal body wall with rosettes. 15–30 µm long. Ventral body wall with grains, 10–30 µm long, that can be perforated as well as simple rosettes, 15–25 µm long. Ventral podia with similar rosettes and few rods which extremities can be sharp or swollen. Dorsal podia with similar assemblage of ossicles as the ventral podia, but also with rods that can take an H-form, 40 µm long.

Processed appearance: Cylindrical shape, slightly tapered at one end. Ventral surface is smooth. Dorsally brown or greyish with small white spots, ventral surface is brown to light-brown. No cuts or small cut across mouth. Common dried size 12–18 cm.

Remarks: Distinguished from *Bohadschia atra* by its lighter colour, more prominent leopard spots, and *B. argus* is not found in the Western Indian Ocean.

Size: Maximum length about 60 cm, commonly to about 36 cm; average fresh weight: 1 800 g (Papua New Guinea and India), 2 000 g (New Caledonia); average fresh length: 35 cm (Australia), 36 cm (Papua New Guinea), 40 cm (New Caledonia and India).

HABITAT AND BIOLOGY: A typical reef species. Generally occurs in 2 to 10 m depth on reef flats and back reef lagoons with clear water. Found mostly on coarse sand near reef structure and can sometimes bury into upper sediments. Populations commonly found at low densities. In the

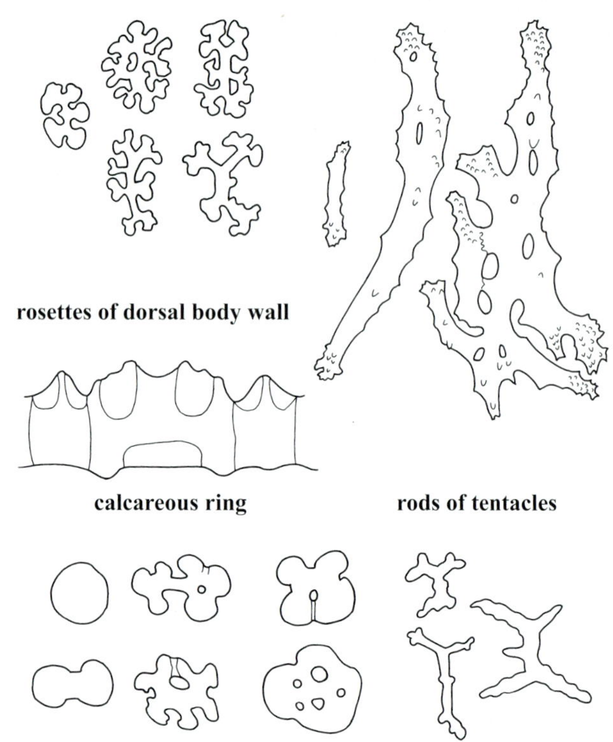

rosettes of dorsal body wall

calcareous ring

rods of tentacles

grains and rosettes of ventral body wall

irregular rods of podia

(after Féral and Cherbonnier, 1986)

western central Pacific, this species prefers barrier reef flats and slopes, or outer lagoons on white sand between 0 and 30 m. In New Caledonia, it spawns in December, while in the Great Barrier Reef it spawns in June.

EXPLOITATION:

Fisheries: Fished in more than a dozen nations throughout the Pacific. Not currently commercially exploited much in Queensland (Australia), or New Caledonia where it has low commercial value. It is part of a subsistence fishery in Wallis and Futuna Islands and Samoa. In Kiribati, fishing for this species boomed between 2000 and 2002 but is now considered depleted. This species is also exported for the aquarium trade from some localities. In Asia, it is of commercial importance in China, Japan, Malaysia, Indonesia, the Philippines and Viet Nam. This species is considered of medium-low commercial importance in China. Fished artisanally in Viet Nam, where it is hand collected, using lead-bombs free-diving and hookah diving, and catches have decreased due to overfishing. In Indonesia, it is heavily fished.

LIVE (photo by: S.W. Purcell)

PROCESSED (photo by: S.W. Purcell)

Regulations: Before a fishery moratorium in Papua New Guinea, fishing for this species was regulated by minimum landing size limits (20 cm live; 10 cm dry) and other regulations. Although exploited little on the Great Barrier Reef, a minimum legal length of 35 cm (fresh) is imposed.

Human consumption: Mostly, the reconstituted body wall (bêche-de-mer) is consumed by Asians. The body wall and/or intestine and gonads may be consumed in traditional diets, e.g. some Pacific islands.

Main market and value: Asia. It is traded at USD15–27 kg^{-1} dried in the Philippines. It is sold at USD20 kg^{-1} dried in Viet Nam and was previously sold at USD6 kg^{-1} in Papua New Guinea. In Fiji, fishers receive USD2–4 per piece fresh. Prices in Guangzhou wholesale markets ranged from USD49 to 63 kg^{-1} dried.

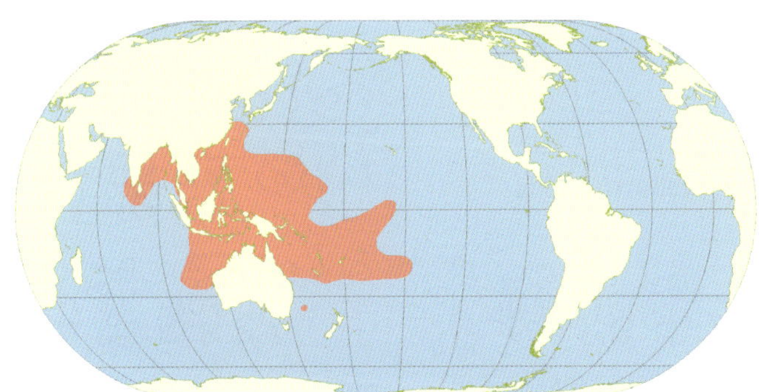

GEOGRAPHICAL DISTRIBUTION: Sri Lanka, Bay of Bengal, East Indies, north Australia, the Philippines, China and southern Japan, south Pacific Islands. In India, it is distributed in the Andamans and Lakshadweep regions and occurs in the far eastern Indian Ocean to French Polynesia in the Pacific.

OTHER USEFUL INFORMATION: Dorsal surface sometimes has foraminifera or small fragments of coral rubble attached to podia.

COMMON NAMES: Falalijaka madarasy and Papiro (Madagascar), Dole (Zanzibar, Tanzania).

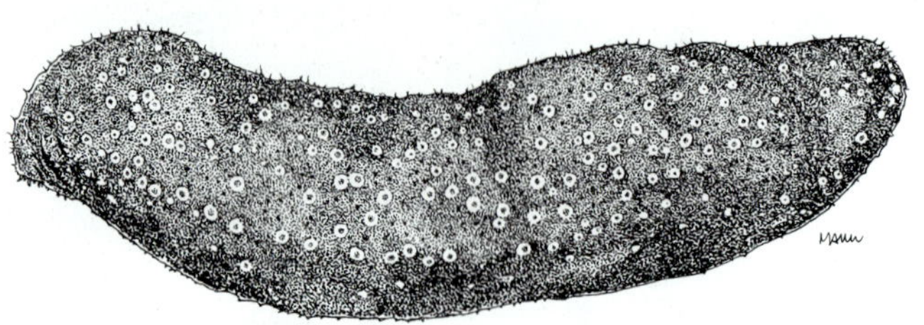

DIAGNOSTIC FEATURES: Colour is deep brown to black dorsally, with numerous brown to red spots surrounding black dorsal papillae. Larger animals may acquire transverse brownsh-red bands. The ventral surface is lighter in colour, tending brown. *Bohadschia atra* is a relatively large holothurian. Body elongated, cylindrical, arched dorsally and flattened ventrally. Dorsal surface has relatively sparse podia and is often covered by sediment. Mouth is ventral with 20 black tentacles. Anus is dorsal and with no anal teeth. Cuvierian tubules are numerous.

Ossicles: Tentacles with rods of various size, dependent on the size of the animal; ranging 80–360 µm in length. Dorsal body wall with relatively simple rosettes. Ventral body wall with similar but somewhat simpler rosettes and grains which can be perforated, all between 20 and 50 µm long. Ventral podia with rosettes similar to those of body wall as wall as straight rods. Dorsal podia with rosettes similar to those of dorsal body wall.

Processed appearance: Cooked and dried animals are elongate and salami shaped, with relatively blunt ends. The dried product is black, and the spots seen in live animals, are not evident. Body texture is relatively smooth. In comparison to other dried animals of similar colour and shape (e.g. *Actinopyga palauensis*, *A. spinea*, *Holothuria atra*), *B. atra* has a dorsally situated anus and no anal teeth.

Remarks: It was recorded as *B. argus* in the Indian Ocean until Massin *et al.* (1999) recognized it as a species new to science. *B. atra* is restricted to the Indian Ocean whereas *B. argus* is largely a Pacific species.

Size: Body length up to 40 cm. Average fresh weight: 500 g; average fresh length: 35 cm.

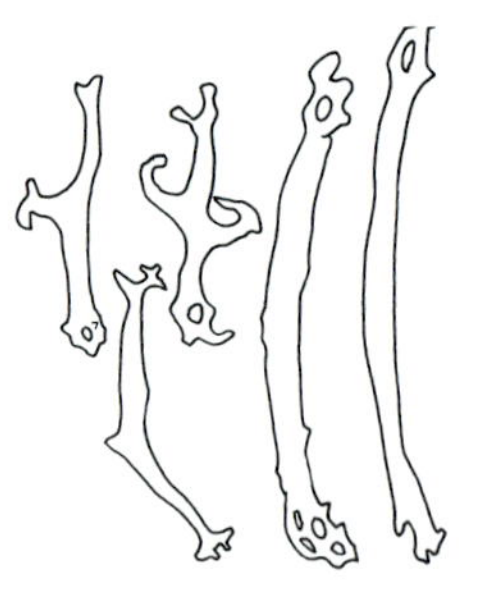

rods of podia

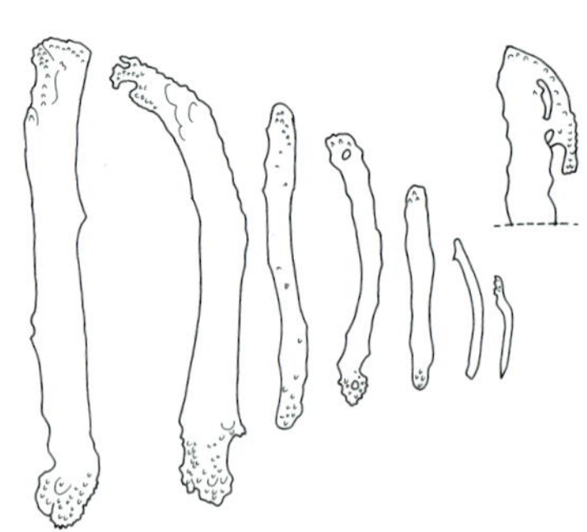

rods of tentacles

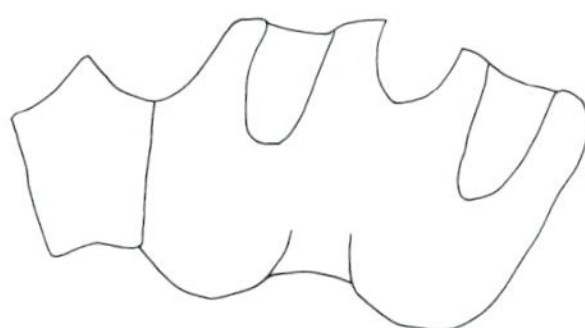

calcareous ring

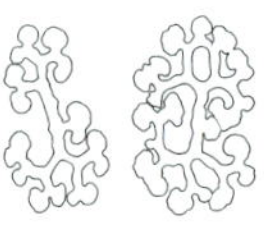

rosettes of dorsal body wall

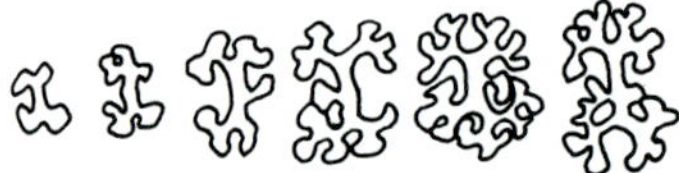

rosettes of dorsal body wall

rosettes of ventral body wall

(source: Samyn, VandenSpiegel and Massin, 2006)

(after Massin *et al.*, 1999)

HABITAT AND BIOLOGY: *B. atra* lives in shallow waters up to 12 m depth, frequently in seagrass beds and sandy areas of coral reefs. Populations are commonly found at low densities.

EXPLOITATION:
Fisheries: It is collected by fishers using free diving and SCUBA in artisanal fisheries. Targeted in multi-species fisheries in many localities of the Western Indian Ocean (e.g. Kenya, Madagascar and Zanzibar [Tanzania]).
Regulations: None except for moratoria on fishing in several countries within the Western Indian Ocean (e.g. Tanzania, Mayotte, Comoros).
Human consumption: Mostly, the reconstituted body wall (bêche-de-mer) is consumed by Asians.
Main market and value: Singapore.

LIVE (photo by: P. Frouin)

PROCESSED (photo by: C. Conand)

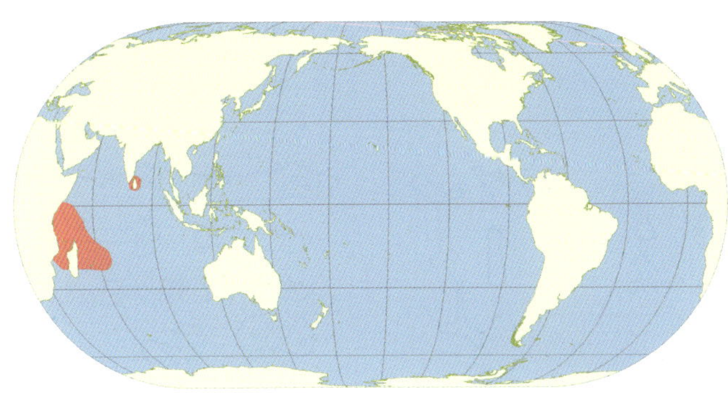

GEOGRAPHICAL DISTRIBUTION:
Southwest region of the Indian Ocean although Conand (2008) states that this species can be found throughout the western Indian Ocean region. Apparently not recorded from the Red Sea or Persian Gulf.

COMMON NAMES: Chalkfish (Papua New Guinea), Brown-spotted sandfish (**FAO**, New Caledonia), Bemangovitse (Madagascar), Finemotu'a (Tonga), Mundra (Fiji).

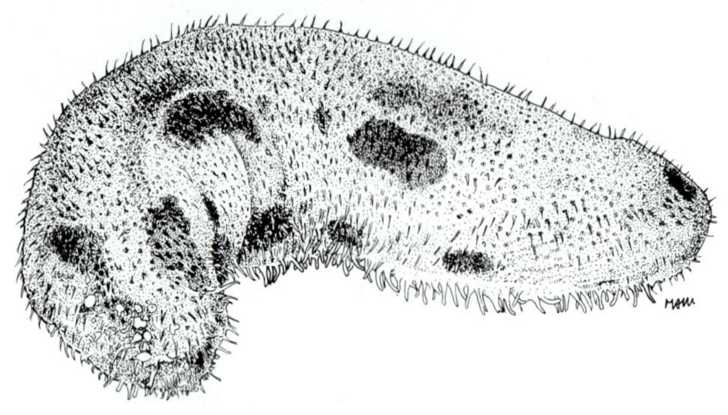

DIAGNOSTIC FEATURES: Normally tan with large brown blotches on the dorsal surface. Ventral surface white to cream colour. *Bohadschia marmorata* is a small to moderate-sized species with a cylindrical body, flattened ventrally and tapering at both ends. It has a very slippery texture. Ventral surface with long slender podia, prominent on the lateral margins. Anus large and nearly dorsal. Juveniles light olive green with darker green blotches, which camouflage them in seagrass beds. May or may not readily eject Cuvierian tubules, depending on the region.

Ossicles: Tentacles with slender rods of various size, up to 220 µm long, and spiny at the extremities. Dorsal body wall with small simple rosettes, 15–20 µm long. Ventral body wall with round, ellipsoid or more irregularly shaped grains, 15–20 µm long and simple rosettes of the same size. Ventral and dorsal tubefeet with few simple rosettes and rods that are mostly little branched distally.

Processed appearance: Bent, narrow cylindrical shape, slightly flattened underside. Dorsal surface is granular, light beige (chalky), underside smooth, black with brown marks. No cuts or small cut across mouth. Common dried size 7–9 cm.

Remarks: Previously synonymised with *B. similis*. Information available for *B. similis* was collated to that of *B. marmorata* in light of genetic analyses by Uthicke, Byrne and Conand (2010).

Size: Maximum length about 26 cm. Average fresh weight 300 g; average fresh length 18 cm.

HABITAT AND BIOLOGY:

Occurs in shallow water rarely deeper than 3 m. Inhabits seagrass beds in muddy-sand sediments, in sheltered or semi-sheltered sites. Predominantly buries in sediments during the day and forages on sediment surface nocturnally. It may be covered with a fine layer of mud. In the western central Pacific region, this species can be found in coastal lagoons and inner reef flats, often found in sandy-muddy substratum. However, in the western Indian Ocean and Africa it prefers the back reef, seagrass beds on sandy bottoms between 0 and 20 m depth.

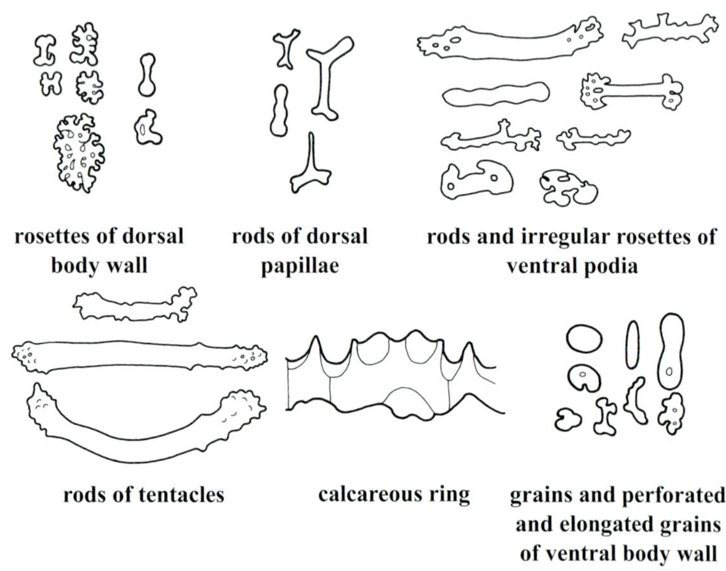

rosettes of dorsal body wall

rods of dorsal papillae

rods and irregular rosettes of ventral podia

rods of tentacles

calcareous ring

grains and perforated and elongated grains of ventral body wall

(after Féral and Cherbonnier, 1986)

This species attains size-at-maturity at 90 g and reproduces annually between February and April.

EXPLOITATION:

Fisheries: It is fished semi-industrially in Mauritius. It is harvested by free-diving and hand collecting. Not currently under commercial exploitation in New Caledonia, where it is considered too low value. This species is commercially exploited in Palau, Micronesia (Federated States of), Tonga, French Polynesia and Fiji. It is also exploited in Indonesia, Malaysia, Thailand, Viet Nam, the Philippines and China. In Seychelles, it is of commercial importance. In northern Kenya, it comprises about 7–8% of the catches. Fishing in Madagascar is limited. In Tanzania, this species is among the most important species despite its low value. Before a moratorium, it was one of the most valuable commercial species in India. In Maldives, a fishery started in 1986 targeting amongst others *B. marmorata*.

LIVE (photo by: S.W. Purcell)

PROCESSED (photo by: E. Aubry SPC)

Regulations: Before a fishery moratorium in Papua New Guinea, fishing for this species was regulated by minimum landing size limits (25 cm live; 7 cm dry) and other regulations. It is not currently of interest to commercial fishers in some countries (e.g. Australia and New Caledonia), so management measures have not yet been established in all fisheries.

Human consumption: Mostly, the reconstituted body wall (bêche-de-mer) is consumed by Asians. It is traded at USD9–22 kg^{-1} dried in the Philippines. In Fiji, it is considered a delicacy or as a protein component in traditional diets, its consumption is important in times of hardship.

Main market and value: China, Hong Kong China SAR. In Fiji, fishers receive USD1.4–2.0 kg^{-1} fresh gutted.

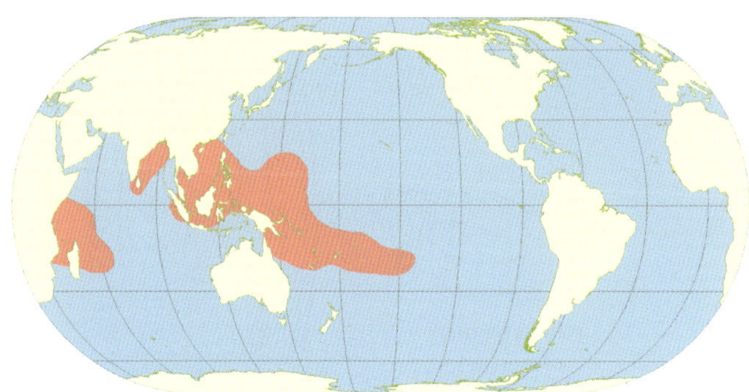

GEOGRAPHICAL DISTRIBUTION:

Mascarene Islands, East Africa, Madagascar, Red Sea, Sri Lanka, Bay of Bengal, East Indies, north Australia, the Philippines, China and southern Japan, South Pacific Islands. Widely distributed in Southeast Asia and the Pacific Islands, where its reported range extends to French Polynesia in the east. Occurs throughout the Indian Ocean to East Africa.

Bohadschia subrubra (Quoy and Gaimard, 1833)

COMMON NAME: Falalyjaka (Madagascar).

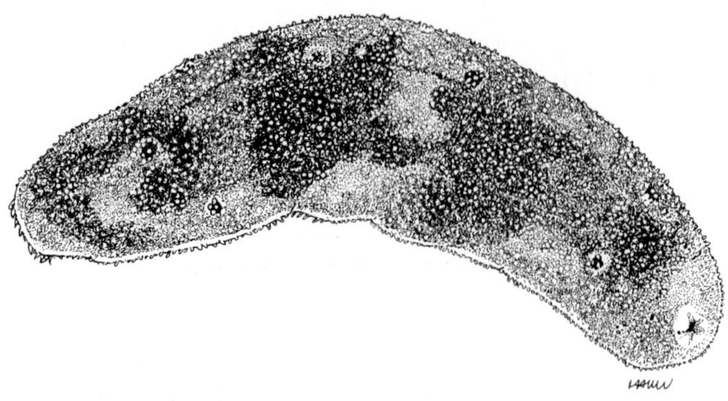

DIAGNOSTIC FEATURES:
This species is a moderately large holothurian with a smooth body texture. Coloration is variable from brown to orange with dark areas dorsally, while the ventral surface is white and distinguished from the rest of the body by a brown line. Body elongated, cylindrical, arched dorsally, flattened ventrally. The dorsal surface is often covered by sediment or seagrass and algal pieces. There are numerous, long, white podia ventrally.

Mouth is ventral with 18 stout white tentacles. Anus is dorsal, and without anal teeth. Cuvierian tubules numerous and large and readily ejected when the animals are disturbed.

Ossicles: Tentacles with rods of various size, 25–540 µm long, mostly spiny, with largest ones forked or with perforated extremities, especially in smaller individuals. Dorsal body wall with rosettes, 20–35 µm long. Ventral body wall with grains of various shapes and rosettes. Ventral podia with rosettes and granules similar to those in the body wall, and with straight non-branched rods, 20–210 µm long. Dorsal podia with rods, 35–230 µm long and rosettes similar to those of the body wall.

Processed appearance description: Elongate shape with blunt, curved ends. Coloration is from orangey-tan to light or dark brown with blotches.

Remarks: Formerly identified as *Bohadschia argus*. Very little is known about this species, as it is exploited and is often found with *B. atra*. A detailed description is given by Massin *et al.* (1999).

Size: Body length up to about 40 cm. Average fresh weight from 500 to 800 g, average fresh length about 35 cm.

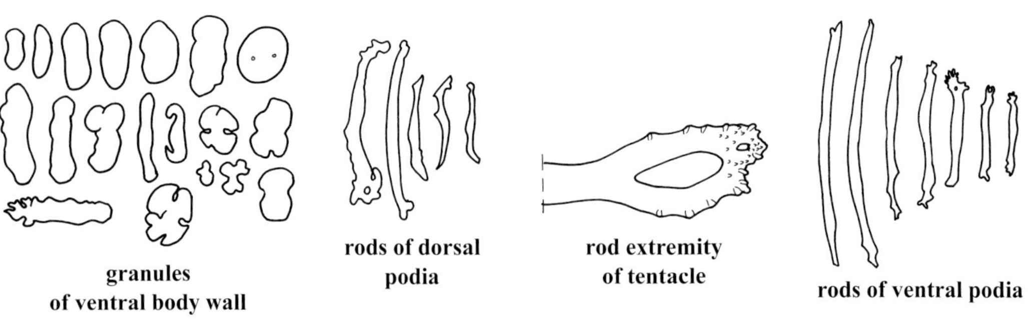

(after Massin *et al.*, 1999)

HABITAT AND BIOLOGY: This species lives in shallow waters, mostly on reef flats, sandy patches, seagrass beds and on coral rubble. It can be found in waters between 0 and 30 m deep. In Kenya, it lives in sandy patches between coral heads in reef lagoons. In the Comoros, it can be found over sand and coral rubble between 5 and 30 m depth. No studies have yet been conducted on its biology.

LIVE (photo by: P. Laboute)

PROCESSED (photo by: H. Eriksson)

EXPLOITATION:

Fisheries: Artisanal. Harvested by snorkel-diving and hand-picking. This species is part of multispecies fisheries, fished together with *Bohadschia atra*, *B. vitiensis*, *Holothuria scabra* and *H. lessoni*.

Regulations: None determined.

Human consumption: Mostly, the reconstituted body wall (bêche-de-mer) is consumed by Asians.

Main market and value: Hong Kong China SAR, Singapore, Taiwan Province of China.

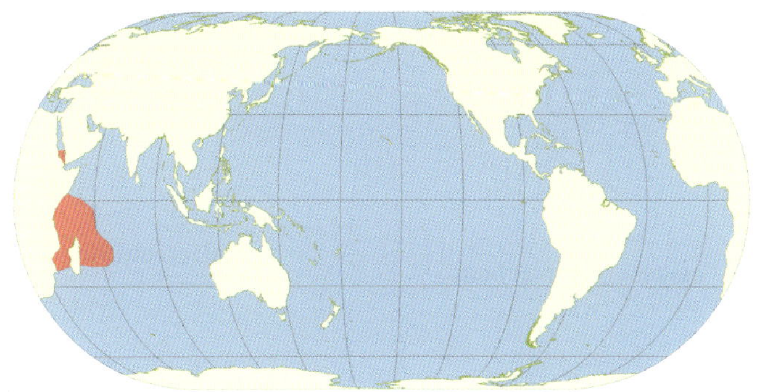

GEOGRAPHICAL DISTRIBUTION: Indian Ocean: Madagascar, Kenya, Seychelles, Mayotte, Mauritius, the Comoros and southern part of Red Sea.

COMMON NAMES: Brown sandfish (**FAO**), Holothurie brune (**FAO**), Nool attai (India), Mangeryfoty, Falalija (Madagascar), Ñoät muû (Viet Nam), Dole (Zanzibar, Tanzania), Pulutan (Philippines), Tekanimnim (Kiribati), Mula (Tonga), Vula (Fiji).

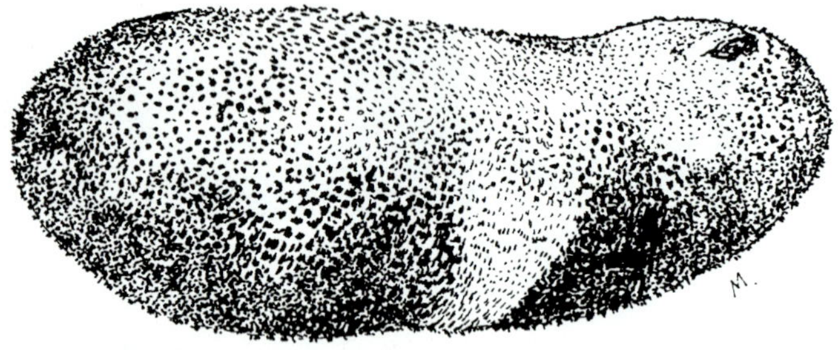

DIAGNOSTIC FEATURES: Colour is variable from cream to yellow-orange to brown with numerous brown spots (around podia) dorsally; whitish ventrally. This species has a stout body, arched highly dorsally, flattened ventrally, often covered by fine sediment. Mouth is ventral with 20 short, yellowish tentacles. The large anus is subdorsal, without teeth. This species has numerous large Cuvierian tubules, which it readily ejects when even slightly disturbed. The juveniles appear similar to small adults.

Ossicles: Tentacles with straight or slightly curved rods. Dorsal body wall with rather stout rosettes, 15–20 µm long, which can occasionally be more elongated. Ventral body wall with grains that can be ovoid, ellipsoid or more irregularly shaped; grains can be perforated, 10–20 µm long. Ventral podia with numerous rods of various shapes, 35–75 µm long. Dorsal podia also with rods, similar in shape and size to the large ones of the ventral podia.

Processed appearance: Cylindrical shape, arched dorsally and moderately flattened ventrally. Brown to brown-black; dorsal surface slightly wrinkled and grainy texture. No cuts or small cut across mouth. Common dried size 12–15 cm.

Remarks: A critical integrative taxonomic review is needed as *Bohadschia vitiensis* is very close, if not identical to: *B. similis, B. tenuissima* and *B. bivittata*.

Size: Maximum length about 40 cm; commonly to about 32 cm. Average fresh weight from 400 and 800 (Réunion and Madagascar) to 1 200 g (Papua New Guinea); average fresh length between 25 and 35 cm.

HABITAT AND BIOLOGY: Found in shallow waters, rarely in depths of more than 20 m. Mostly in sheltered coastal lagoons and inner reef flats on sand or occasionally among rubble and coral patches. This species can also be abundant in sandy-muddy sediments where it buries during parts of the day.

It reproduces annually during the dry season. In Papua New Guinea, it reproduces sexually in December, while in Palau it reproduces in July and August. In eastern Africa and the Indian Ocean, it reproduces during the warm season.

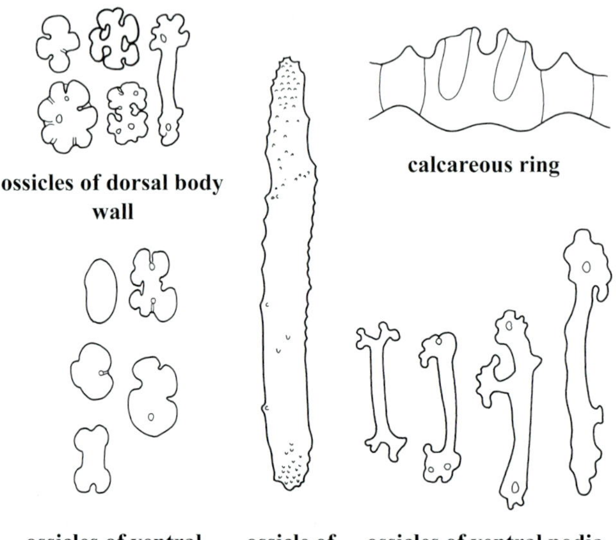

(after Cherbonnier, 1988)

EXPLOITATION:

Fisheries: Harvested in artisanal fisheries by hand collecting and free diving (Réunion and Madagascar), occasionally using lead-bombs (Papua New Guinea), or using hookah diving gear (Viet Nam). This species is commercially exploited in most countries of the Indo-Pacific as far east as French Polynesia. It is fished for subsistence use in Samoa and Fiji. In Kiribati, a multispecies fishery that included *B. vitiensis* boomed from 2000 to 2002, but is now considered overfished. In Madagascar and New Caledonia, harvesting of this species is minimal. Stocks in Australia are probably unfished.

Regulations: Before a fishery moratorium in Papua New Guinea, fishing for this species was regulated by minimum landing size limits (20 cm live; 10 cm dry) and other regulations. In Samoa, exportation of sea cucumbers is banned to safeguard remaining stocks of species like *B. vitiensis* for local consumption.

Human consumption: Mostly, the reconstituted body wall (bêche-de-mer) is consumed by Asians. It is considered a delicacy or as a protein component in traditional diets or eaten in times of hardship.

Main market and value: Hong Kong China SAR, Singapore, Taiwan Province of China, Asia. In Viet Nam, Ho Chi Minh City for further export to Chinese markets. It is sold by fishers at USD6 kg⁻¹ dried. In Fiji, fishers receive USD0.4–1.4 per piece fresh. Prices in Hong Kong China SAR retail markets ranged from USD103 to 167 kg⁻¹. Wholesale prices in Guangzhou were up to USD49 kg⁻¹ dried.

LIVE (photo by: S.W. Purcell)

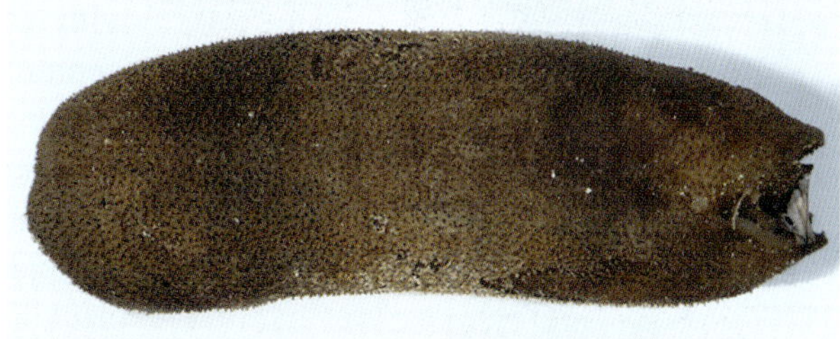

PROCESSED (photo by: S.W. Purcell)

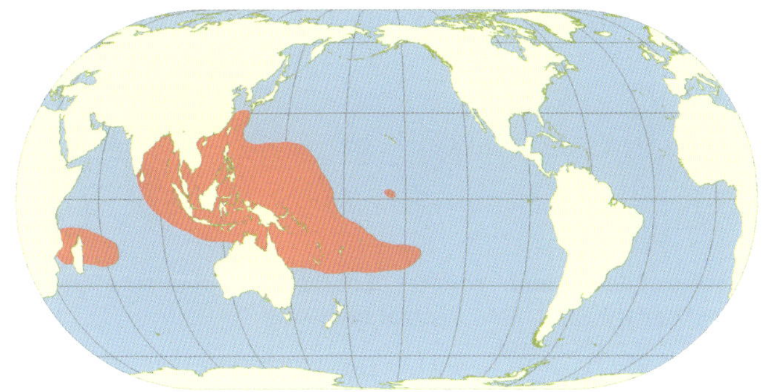

GEOGRAPHICAL DISTRIBUTION:

Widely distributed in the Indo-Pacific, being reported as far east as French Polynesia and west to Madagascar and eastern Africa.

Pearsonothuria graeffei (Semper, 1868)

COMMON NAMES: Flowerfish (India, Papua New Guinea, Viet Nam), Orange fish (Egypt), Blackspotted sea cucumber (**FAO**), Ñoät daûi, Ñoät daûi ñaù or Daâu ñaù (Viet Nam), Shoab (Egypt), Nool attai (India), Trompa, Piña, Mani-mani or Bulaklak (Philippines), Zanga somotse, Tigre (Madagascar).

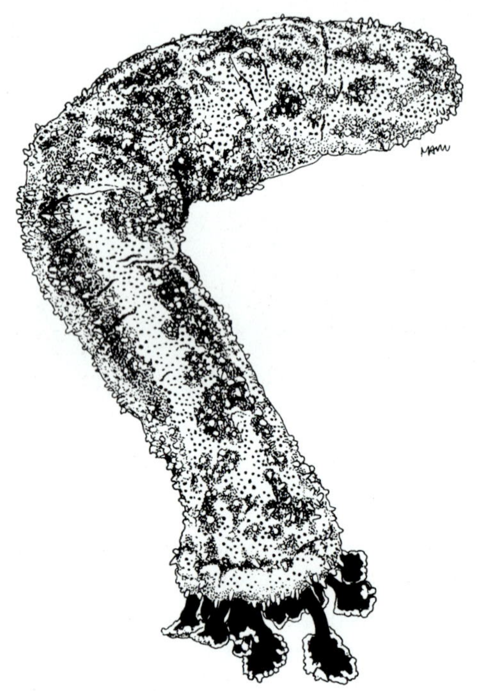

DIAGNOSTIC FEATURES: Coloration is cream to tan with numerous large brown patches and with fine dark speckling. Body relatively elongated, cylindrical, and with numerous transverse folds. It is flattened somewhat ventrally. Moderately long (3–5 mm) conical papillae with white tips are scattered over the dorsal surface. Ventral surface with three bands of numerous, long, brown podia. The mouth is ventral with 23 –28 black tentacles with distinctive white edge. Anus terminal without teeth or papillae. Cuvierian tubules present, but not ejected. Juvenile mimic nudibranchs; white with black lines and large, conical, yellow papillae.

Ossicles: Tentacles with rods, which can take a rosette form, 20–90 μm long. Dorsal and ventral body wall with the same type of knobbed pseudo-tables, 30–65 μm long, and rosettes, 20–50 μm long. Ventral and dorsal podia with very complex rosettes that resemble those of the body wall.

Processed appearance: Elongated with a rectangular cross-section. Black to black-brown. Dorsal surface is rough, while the ventral surface is grainy. No cuts or small cut across mouth. Common dried size 15 cm.

Remarks: Previously misnamed as *Bohadschia graeffei* or *B. drachi*.

Size: Maximum length about 45 cm; commonly to about 35 cm. Average fresh weight: 130 g (Philippines), 300 g (Réunion), 500 g (Egypt), 600 g (India) and 700 g (Papua New Guinea); average fresh length: 17 cm (Philippines), 30 cm (Egypt, India, Réunion, Viet Nam) and 35 cm (Papua New Guinea).

HABITAT AND BIOLOGY: Inhabits hard surfaces on coral reefs, commonly broken dead coral or reef pavement. Lives on reef slopes, commonly with live coral, in shallow waters between 0 and 25 m. It often feeds during

pseudo-tables of dorsal and ventral body wall   rosettes of dorsal and ventral body wall   rosettes of podia

ossicles of anal body wall   ossicles of tentacles   calcareous ring

(after Cherbonnier and Féral, 1984)

the day and night on detritus on hard surfaces. This species does not bury. Little is known of its biology. This species is known to host the pearl fish *Carapus boraborensis*. On the Great Barrier Reef (Australia), it reproduces between November and February.

EXPLOITATION:

Fisheries: Generally fished artisanally. Harvested by hand collecting (Philippines, Papua New Guinea), free diving and lead-bombs (Papua New Guinea), SCUBA diving (Egypt), and hookah diving (Viet Nam).

This species is commercially exploited in Guam, Micronesia (Federated States of), Kiribati, Papua New Guinea, Solomon Islands, Fiji, Malaysia, Madagascar, the Philippines (referred to as *B. graeffei*) and Indonesia.

Regulations: Before a fishery moratorium in Papua New Guinea, fishing for this species was regulated by a fishing season, a (global) total allowable catch (TAC), gear restrictions and permits for storage and export.

Human consumption: Mostly, the reconstituted body wall (bêche-de-mer) is consumed by Asians. Also used in traditional medicine (Egypt).

Main market and value: In the Philippines, it is sold for USD2–5 kg^{-1} dried. In Fiji, fishers receive USD0.3–0.6 kg^{-1} fresh gutted.

LIVE (photo by: S.W. Purcell)

PROCESSED (photo by: J. Akamine)

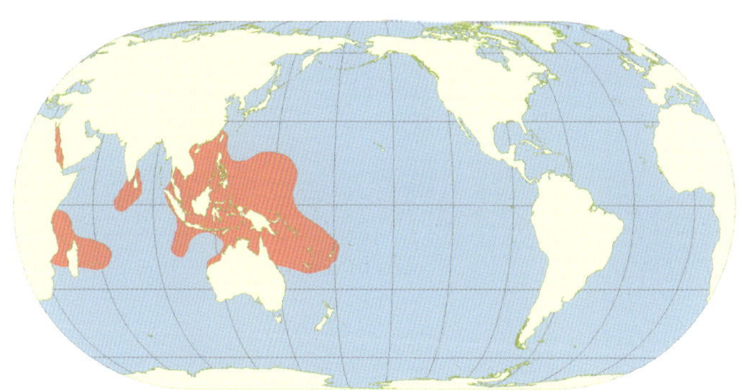

GEOGRAPHICAL DISTRIBUTION:

Widespread in the Indo-Pacific, including the Red Sea and the Gulf of Aqaba, Maldives, India, the Philippines, and South Pacific Islands as far east as Fiji.

Holothuria arenicola Semper, 1868

COMMON NAMES: Tripang kappallah poetih (Indonesia).

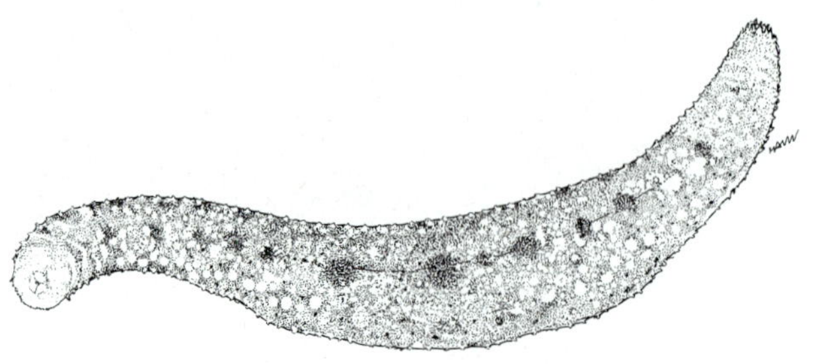

DIAGNOSTIC FEATURES: Coloration is cream to rusty tan. Some individuals are quite orange, becoming whitish towards either the mouth or anus. Ventral surface is yellowish-white with two rows of relatively large, dark brown, spots. *Holothuria arenicola* is a small species. The body wall is relatively thin but is very rough to the touch. No Cuvierian tubules.

Ossicles: Tentacles with rods, spiny at their extremities; largest ones have distal ends perforated and/or forked, 100–200 µm long. Dorsal and ventral body wall with buttons and tables of the same kind and size: tables very small, 40–55 µm across, flat, often reduced to the smooth-edged disc, perforated by 4 large central holes and 0–4 small peripheral holes, with very short pillars that end in a few spines or a small crown of spines; buttons very regular, 40–50 µm long, with 3 pairs of holes. Ventral podia with buttons, tables and rods, up to 180 µm long. Dorsal podia with tables, buttons and rods similar to those of ventral podia, but large buttons, 50–225 µm long, with 3–10 pairs of holes, are abundant.

Processed appearance: Dried animals are light-brown to medium-brown and often with calcareous deposits remaining on the body wall. The body tapers at both ends.

Remarks: This species is part of a complex of species, with unrecognized forms included. Even some records from the East African coast are potentially misidentifications with other species, such as *H. strigosa*.

Size: Maximum length about 30 cm; commonly to about 10 cm. Average fresh weight probably <100 g.

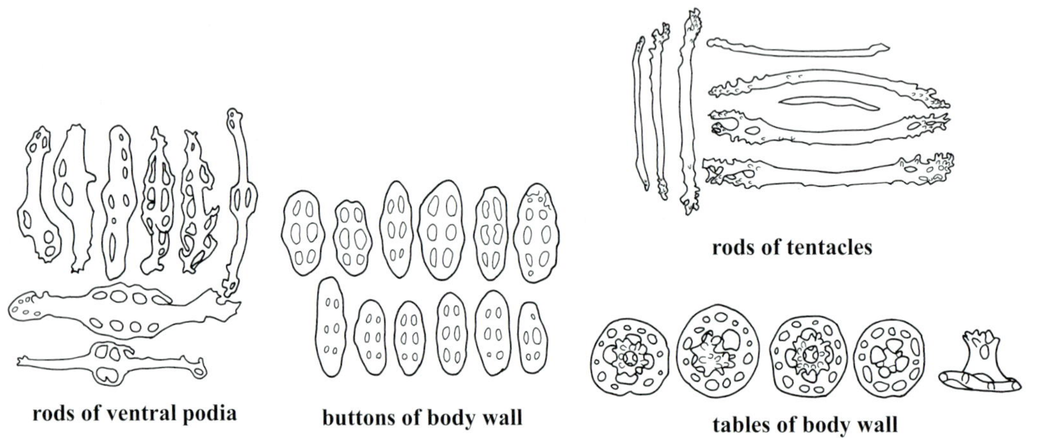

rods of ventral podia   buttons of body wall   rods of tentacles   tables of body wall

(after Massin, 1996)

">

HABITAT AND BIOLOGY: Abundant in intertidal and shallow areas but can also be found in deeper waters. It can be found under stones, in coral debris and on sand flats. Specimens have been found buried in *Thalassia* seagrass beds in 3 m of water. In Honduras, it buries in sandy substrata and seagrass beds but it has also been found under rubble and in dead conch shells. This sea cucumber can form conical mounds where it buries. This species ingests surface and subsurface sediments using a funnel that ends 15 to 20 cm below the surface.

LIVE (photo by: A.M. Kerr)

PROCESSED (photo by: S.W. Purcell)

EXPLOITATION:

Fisheries: This species is believed to be fished in China, Madagascar and Egypt. The scale of fishing is mostly artisanal.

Regulations: Management regulations are generally lacking in countries in which it is fished.

Human consumption: Poorly known.

Main market and value: It is a low-value species. Retail prices in Hong Kong China SAR were up to USD2 kg^{-1} dried.

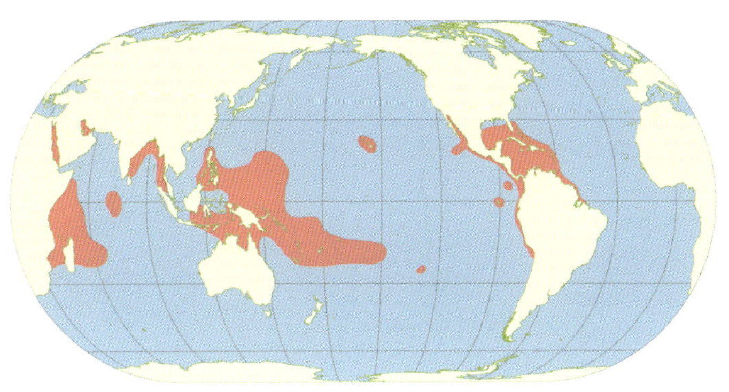

GEOGRAPHICAL DISTRIBUTION: This species is believed to be found at some localities in the Western Pacific, parts of Asia, and the Indian Ocean, including the Red Sea and the Comoros. Reported along the Pacific coast of Central America. This species is reported from the Caribbean and Brazil, but those sightings probably represent a different species.

COMMON NAMES: Lollyfish (**FAO**), Barbara (Mauritius), Stylo noir (Madagascar), Kuchii attai (India), Sherman (Egypt), Ñia ñen, Ñia maùu (south Viet Nam), Black beauty and Mani (Philippines), Black lollyfish (Africa and Indian Ocean region), Lega (Eritrea), Kichupa (Zanzibar, Tanzania), Loli (Tonga), Tentabanebane (Kiribati), Loliloli (Fiji).

DIAGNOSTIC FEATURES: Uniformly black. Body is commonly covered with medium-grain sand, with characteristic bare circles in two rows along the dorsal surface. This species has three morphs: a small morph (common) that is smooth and covered with sand; a morph on reef crests that has a rippled dorsal surface; and a large morph with little sand on its body, occurring more commonly in deeper waters. The three morphs are genetically indistinguishable. Podia on the dorsal surface are small and sparse. Tentacles are black. Anus is terminal, without teeth or papillae. Cuvierian tubules absent. This species can also be distinguished by the reddish dye released from its body wall when rubbed.

Ossicles: Tentacles with simple slender rods of various size. Dorsal and ventral body wall with same type of tables and rosettes. Tables of ventral body wall with larger, more spinose disc, up to 60 µm across, than those of dorsal body wall. Table disc perforated by four central holes; spire ending in a Maltese cross. Rosettes simple, 20–25 µm long, more abundant dorsally than ventrally. Podia with pseudo-plates, 75–100 µm long, and rosettes of similar size as those in the body wall.

Processed appearance: Narrow cylindrical shape. Entire body surface smooth and black but specimens of the larger reef variety have transverse wrinkles across the dorsal surface. No cuts or small cut across mouth. Common dried size 5–12 cm.

Size: Maximum length about 45 cm (large morph). Average fresh weight: 200 g (Papua New Guinea and India), 300 g (Egypt), 335 g (Viet Nam), 400 g (Mauritius); average fresh length: 15 cm (Mauritius), 20 cm (Papua New Guinea and India), 23 cm (Viet Nam), 30 cm (Egypt).

HABITAT AND BIOLOGY: Inhabits the inner and outer flats, back reefs, shallow lagoons, sand-mud and rubble, and seagrass beds between 0 and 20 m. In Mauritius, it can be found in areas with the calcareous algae, *Halimeda* sp. In Mauritius, this species reaches size-at-maturity at 80 g drained weight. Its reproductive season is between February and April. In Fiji, it reproduces between September and December, while in the Great Barrier Reef (Australia) in January, May–June, and November–December. This species can reproduce asexually by fission in natural conditions, which seems to occur seasonally. In Réunion, spawning occurs in the warm season while fission occurs in the cool season. Halimeda sp.

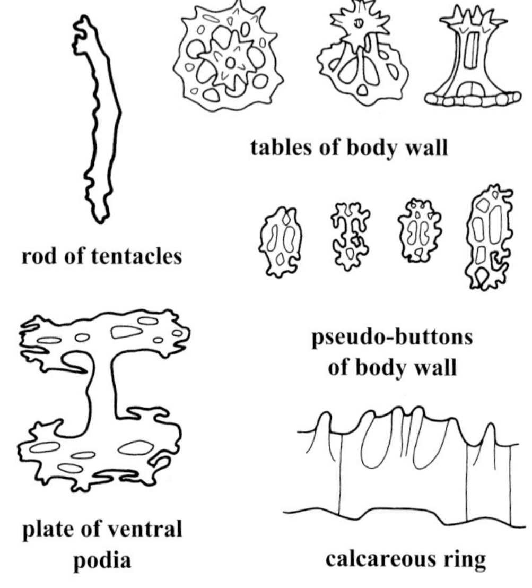

tables of body wall

rod of tentacles

pseudo-buttons of body wall

plate of ventral podia

calcareous ring

(after Cherbonnier, 1980)

EXPLOITATION:

Fisheries: Fished artisanally (e.g. Viet Nam, Kiribati), semi-industrially (Mauritius), and industrially (Egypt). Harvested by hand collecting, free-diving, hookah diving, and SCUBA diving. It is, or has previously been, harvested in at least 20 countries and islands States in the western central Pacific. Harvested for subsistence in Guam, Nauru, Samoa, Cook Islands, Nuie and French Polynesia. It is of commercial importance in China, Japan, Malaysia, Thailand, Indonesia, the Philippines and Viet Nam. Commercially important in Tanzania, Mauritius and Eritrea. There is some harvesting of this species in Sri Lanka, Egypt, Madagascar, Mozambique and Seychelles. In the Galápagos Islands (Ecuador), it is fished illegally.

Regulations: Before a fishery moratorium in Papua New Guinea, fishing for this species was regulated by minimum landing size limits (30 cm live; 15 cm dry) and other regulations. Although seldom fished in Australia, a minimum live size limit is set at 15 cm in Torres Strait, the Northern Territory and Western Australia, and 20 cm on the Great Barrier Reef. In Maldives, the minimum live size limit of this species is 15 cm.

Human consumption: In few Pacific Island nations, the body wall, intestines and/or gonads are consumed in traditional diets or in times of hardship. More often, it is dried and exported for consumption, predominantly by Asians.

Main market and value: China and Hong Kong China SAR. Ho Chi Minh City in Viet Nam for further export to the Chinese market. It has been traded at USD4–20 kg^{-1} dried in the Philippines. In Papua New Guinea it was previously sold at USD2.5 kg^{-1} dried. In Fiji, fishers receive USD0.6–1.4 kg^{-1} fresh gutted. Retail prices in Hong Kong China SAR were up to USD210 kg^{-1}. Wholesale prices in Guangzhou were up to USD63 kg^{-1} dried.

LIVE (photo by: S.W. Purcell)

PROCESSED (photo by: S.W. Purcell)

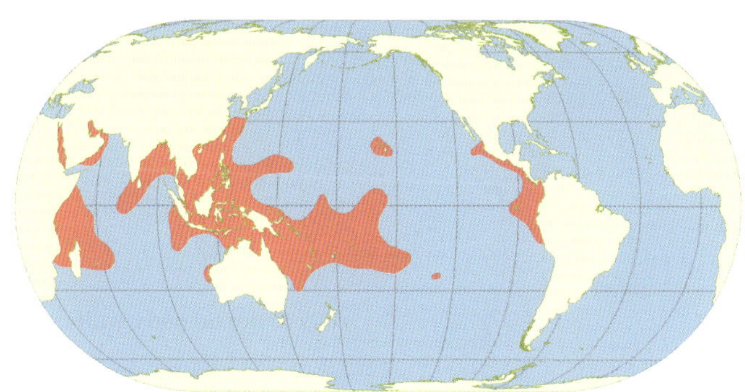

GEOGRAPHICAL DISTRIBUTION:

Widespread in the Indo-Pacific. This species is found at Mascarene Islands, East Africa, Madagascar, Red Sea, southeast Arabia, Persian Gulf, Maldives, Sri Lanka, Bay of Bengal, India, North Australia, the Philippines, China and southern Japan, South Sea Islands, Hawaiian Islands. It can be found in the islands in the central and eastern tropical Pacific, including Coco and Galápagos islands, Panama region, Clipperton Island and Mexico.

Holothuria cinerascens (Brandt, 1835)

COMMON NAME: Zanga fleur (Madagascar).

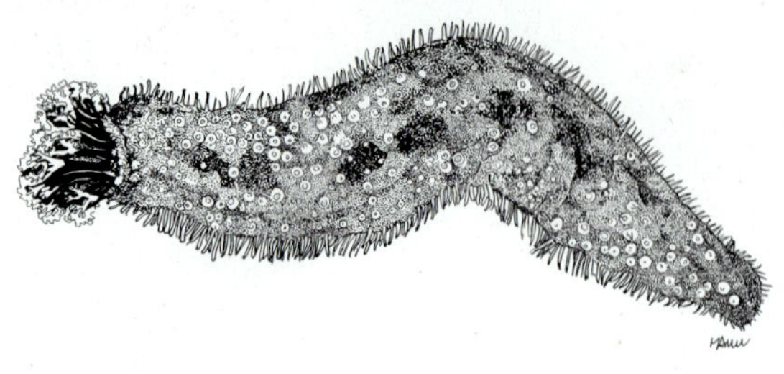

DIAGNOSTIC FEATURES: Coloration is rusty to dark brown, with podia and tentacles lighter. Cylindrical body. The relatively long podia are abundant ventrally and numerous dorsally. Terminal mouth with 20 dendro-peltate tentacles, well developed. The terminal anus has 2 or 3 small papillae. This species does not have Cuvierian tubules.

Ossicles: Tentacles with rods, 60–140 µm long, finely rugose at the sides. Dorsal and ventral body wall with similar tables and rods. Tables more numerous in the dorsal than in the ventral body wall. Tables with discs 35–55 µm across, smooth to slightly spinose rim, perforated by 4 central holes; spire ending in a wide Maltese cross. Rods rugose, more so dorsally, 65–100 µm long. Ventral and dorsal podia with similar ossicles to those in the body wall, but dorsal ones also with perforated plates, up to 120 µm long.

Processed appearance: Grey in colour with brown lines ventrally corresponding to rows of podia in the fresh animal. The dried product is shaped like small wooden sticks, and the buccal tentacles are extended, which is unusual in processed sea cucumbers.

Remarks: This species extends its tentacles out from sand and reef crevices where it seeks refuge.

Size: Maximum length about 16 cm. Average length is about 10 cm.

HABITAT AND BIOLOGY: In the Africa and Indian Ocean region this species lives in the outer reef over hard substratum generally between 0 and 3 m, but believed to be found at up to 20 m depth. It can be found over rocky bottoms in crevices with strong wave action where it suspension feeds organic particles from the water column. In New Caledonia, it lives buried in coral sand with the tentacles extended for suspension feeding in sites with currents or wave action. In the Comoros, this species is found in the intertidal region. Similarly, in the Pacific it can be found on shallow reef flats. In Taiwan Province of China, it reproduces between April and June.

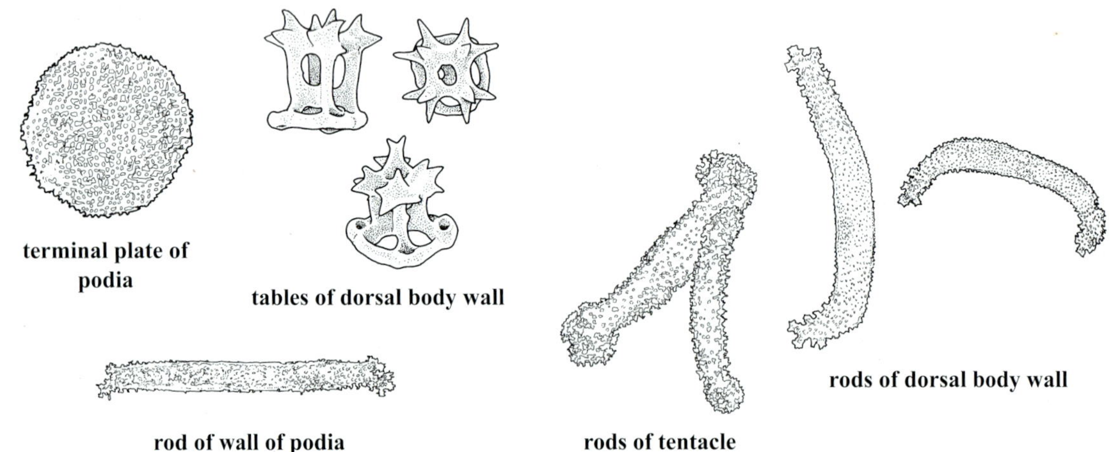

terminal plate of podia

tables of dorsal body wall

rods of dorsal body wall

rod of wall of podia

rods of tentacle

(source: Samyn, VandenSpiegel and Massin, 2006)

EXPLOITATION:

Fisheries: This species is fished in Guam and in the southern Cook Islands in subsistence fisheries where its gonads are eaten. It is of commercial importance in China. It is occasionally fished in Madagascar.

Regulations: There are few regulations pertaining to the harvesting of this species.

Human consumption: The gonads of this species are eaten in subsistence fisheries. The body wall is also consumed by Asians.

Main market and value: Unknown, but considered a low-value species.

LIVE (photo by: P. Bourjon)

PROCESSED (photo by: C. Conand)

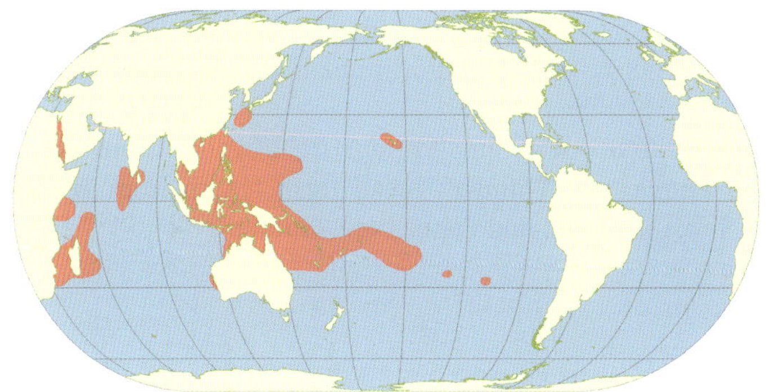

GEOGRAPHICAL DISTRIBUTION:

East Africa and Indian Ocean, including the Red Sea, Maldives, India and Indonesia. Distribution continues into to the South China Sea, the Philippine Sea and Pacific Ocean, including China, the Philippines, Guam, Australia, Hawaii, Japan, New Caledonia, Cook Islands and as far east as Easter Island.

COMMON NAMES: Snakefish (**FAO**), Holothurie serpent (**FAO**), Bat uwak, Tambor and Patola white (Philippines), Kichupa (Zanzibar, Tanzania), Te'epupulu maka (Tonga).

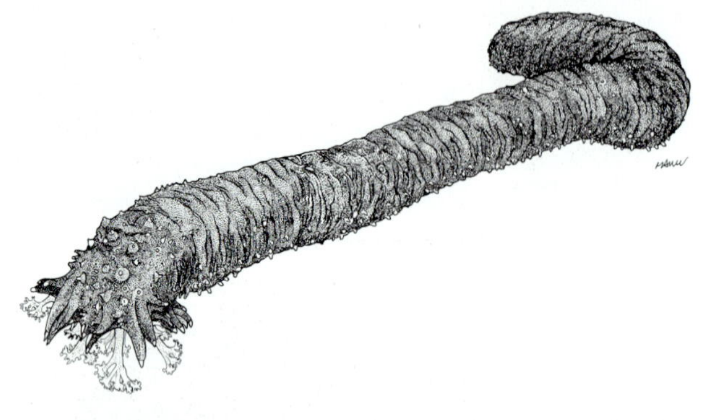

DIAGNOSTIC FEATURES: Dark grey to black in colour, with distinctive yellow-tipped papillae, especially around the ventral margins and around the mouth. A very elongate holothurian, cylindrical and narrowing near anterior end. Stiffened tegument, rough to the touch. The large mouth, which fans laterally, has large pale yellow tentacles, which also distinguish it easily from *Holothuria leucospilota* (which has black tentacles). Cuvierian tubules absent.

Ossicles: Tentacles with slightly curved and distally spinose rods, 40–165 µm long. Dorsal and ventral body wall with similar tables and buttons. Tables with disc 60–80 µm across, perforated by a single central and up to 12 peripheral holes, edge spinose and often turned upwards to give a so called 'cup and saucer' appearance; spire low ending in a narrow crown of spines. Buttons are rare, and they have 3–5 pairs of holes and quite irregular in appearance. Podia with tables as in body wall and, in addition, perforated plates, 35–140 µm long, and rods with enlarged and often perforated extremities, 50–110 µm long.

Processed appearance: Long irregular skinny shape, clearly tapered at the anterior end. Brown body covered with tiny whitish bumps. Small cut across mouth and/or in the body middle. Common size 12–18 cm.

Size: Maximum length about 60 cm, commonly to about 40 cm. Average fresh weight from 140 g (Philippines) to 300 g (New Caledonia and Papua New Guinea); mean fresh length from 18 cm (Philippines), 26 cm (Indonesia), to 40 cm (New Caledonia and Papua New Guinea).

HABITAT AND BIOLOGY: Occurs in coastal and sheltered waters on reef flats, muddy-sand bays and reef lagoons and patch reef systems. Generally found in shallow water 0 to 8 m depth, but occasionally up to 15–25 m depth. Characteristically, this species shelters under boulders and at the edges of patch reefs, where it lodges its posterior end in hard structures, and ventures out far with its anterior end to feed on the surface of soft sediments. It reaches a size-at-maturity at 13 to 15 cm. It has an annual reproductive event.

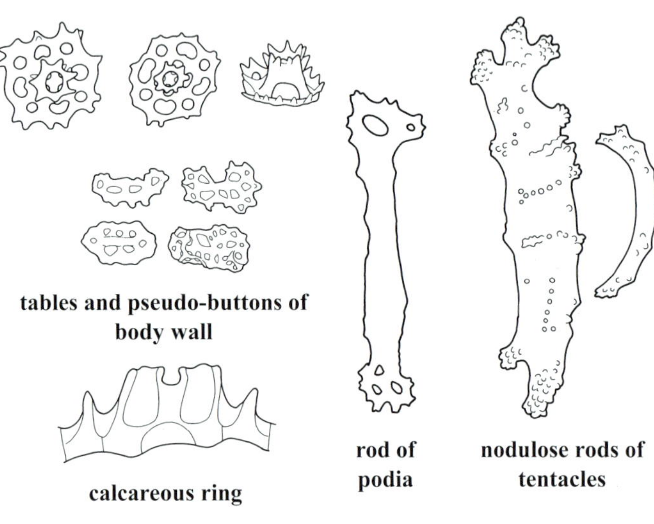

(after Féral and Cherbonnier, 1986)

EXPLOITATION:

Fisheries: *H. coluber* is harvested in artisanal fisheries in much of its range. In the Philippines, Papua New Guinea, New Caledonia, Fiji and Tonga, this species is collected by hand by gleaners on reef flats, but also harvested by free diving in shallow waters. It is fished in many countries within its distribution in the western central Pacific. It is heavily exploited in Indonesia and the Philippines and sometimes fished in Madagascar. The Philippines reportedly export this species to China to be used as a fertilizer.

Regulations: Before a fishery moratorium in Papua New Guinea, fishing for this species was regulated by a fishing season, a (global) TAC, gear restrictions and permits for storage and export.

Human consumption: Mostly, the reconstituted body wall (bêche-de-mer) is consumed by Asians.

Main market and value: Asia. In the Philippines, it is sold at USD4–20 kg^{-1} dried. In Fiji, fishers receive USD1.4–2.5 kg^{-1} fresh gutted. Wholesale prices in Guangzhou were up to USD38 kg^{-1}.

LIVE (photo by: S.W. Purcell)

PROCESSED (photo by: J. Akamine)

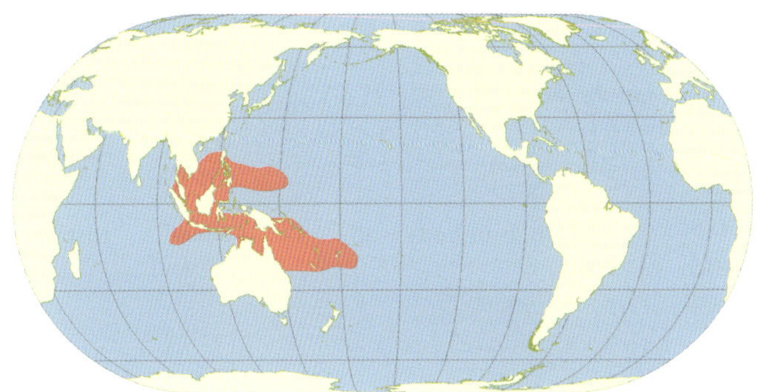

GEOGRAPHICAL DISTRIBUTION:

Common in the Pacific, as far east as Tonga in the south and in the northern Pacific to Pohnpei and Kosrae in Micronesia. It is also widely distributed in Southeast Asia, including Viet Nam, Indonesia, Malaysia and Timor Sea. Apparently absent in much of the Indian Ocean, but has been recorded from western Indonesia and Cocos (Keeling) Islands.

COMMON NAMES: Pinkfish (**FAO**), Trépang rose (**FAO**), Kadal Attai (India), Saâu gai (south Viet Nam), Red beauty, Red-black, Hotdog (Philippines), Pink lollyfish (Africa and Indian Ocean region), Abu sanduk tina (Eritrea), Stylo rouge (Madagascar), Cera (Indonesia), Loli kula (Tonga), Tenautonga (Kiribati), Dri-damu (Fiji).

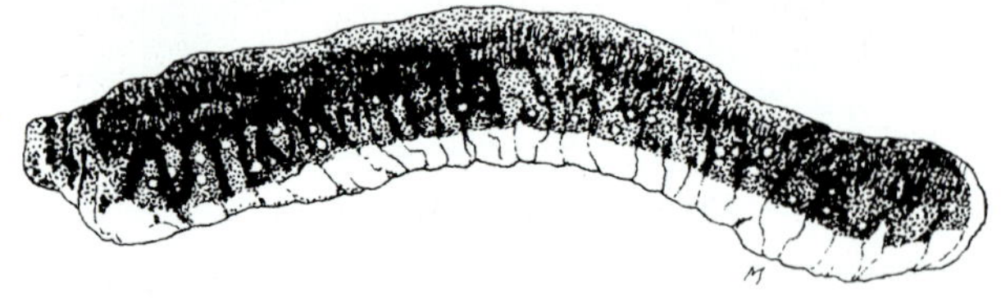

DIAGNOSTIC FEATURES: Dark grey, chocolate brown or black dorsally, fading laterally to pink or whitish pink ventrally. Ventral surface with small dark spots. A small to medium-sized, sausage-shaped holothurian. Body subcylindrical. Tegument somewhat rough with sparse papillae on dorsal surface. Ventral podia are short but stout, numerous and light coloured. Anus terminal and Cuvierian tubules absent.

Ossicles: Tentacles with curved rods that have enlarged spiny extremities, 70–180 µm long. Dorsal and ventral body wall with similar tables and button-like rosettes. Tables with disc greatly reduced, on average 35 µm across, perforated by 1 central hole; spire ending in a Maltese cross. Button-like rosettes perforated by 4–10 uneven holes and with uneven rim, 30–70 µm long. Ventral podia with perforated plates, 100–140 µm long, and shorter rods. Dorsal podia with large rods that can have few perforations, 135 µm long on average.

Processed appearance: Narrow cylindrical shape, slightly flattened ventrally. Dorsal surface with small wrinkles, dark brown; ventral surface is smoother, light to medium brown. No cuts or small cut across mouth. Common dried size 10–14 cm.

Size: Maximum length 38 cm; commonly to about 24 cm. Average fresh weight 200 g; average fresh length 20 cm.

HABITAT AND BIOLOGY:

Found mostly on silty-sand or sand mixed with coral rubble. Occupies semi-sheltered reef habitats, namely reef flats and lagoon patch reefs near the coast from 0 to 20 m depth. Also found in seagrass beds and sometimes on hard reef surfaces.

Asexual reproduction by fission is annual, but the sexual reproduction cycle is uncorrelated and appears continuous. In southern Viet Nam, transverse fission has been recorded in June at the beginning of the rainy season. On the Great Barrier Reef, this species reproduces sexually between December and January.

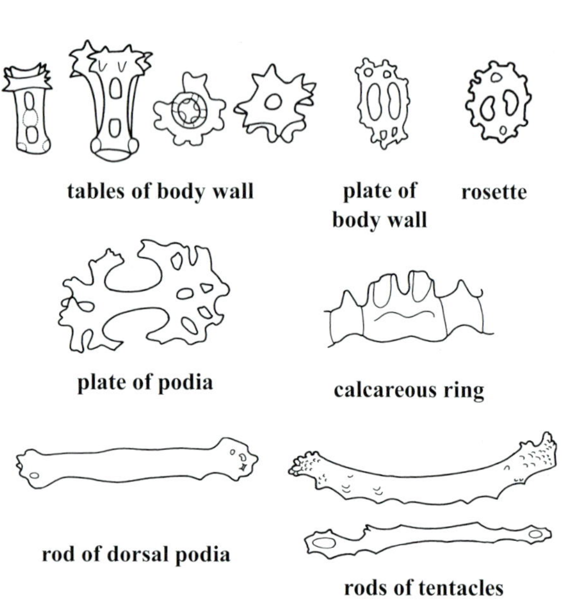

(after Cherbonnier, 1980)

EXPLOITATION:

Fisheries: It is harvested in artisanal fisheries in much of its range in the Indo-Pacific. Prior to a moratorium, harvesting in Papua New Guinea involved hand collection, free diving and use of lead-bombs. Minor harvesting in New Caledonia is done by hand by gleaning on reef flats at low tide or skin diving in shallow waters. In Viet Nam, this species is gathered using hookah diving. In Asia, this species is fished in China, Japan, Malaysia, Thailand, Indonesia, the Philippines and Viet Nam. In Indonesia and the Philippines, it is heavily exploited. This species is of low commercial importance in Kenya and Madagascar. It is not of commercial importance in Seychelles, New Caledonia and Australia. It is sometimes collected for the aquarium trade.

LIVE (photo by: S.W. Purcell)

PROCESSED (photo by: S.W. Purcell)

Regulations: In New Caledonia, there is a prohibition for the use of compressed air apparatus, fishers must be licensed and there are no-take reserves. Before a fishery moratorium in Papua New Guinea, fishing for this species was regulated by minimum landing size limits (25 cm live; 10 cm dry), a fishing season, a TAC, gear restrictions and permits for storage and export.

Human consumption: Mostly, the reconstituted body wall (bêche-de-mer) is consumed by Asians.

Main market and value: Asia. It is a low-value species. In Viet Nam, Ho Chi Minh City for further exports to Chinese markets. It has been traded at USD4–20 kg^{-1} dried in the Philippines.

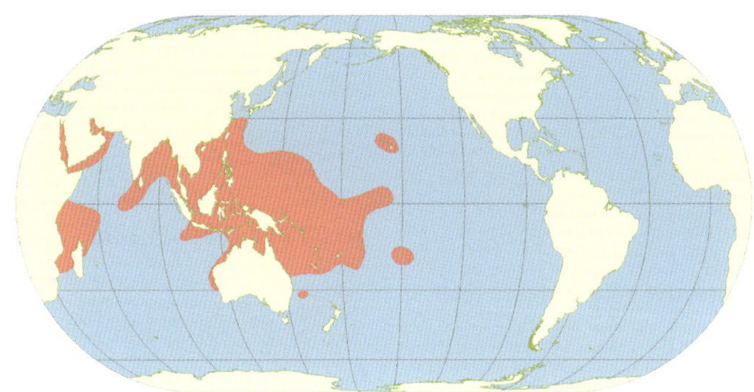

GEOGRAPHICAL DISTRIBUTION:

East Africa, Madagascar, Red Sea, southeast Arabia, Sri Lanka, Bay of Bengal, East Indies, North Australia, the Philippines, China and southern Japan, South Sea Islands. In India, this species is distributed in the Gulf of Mannar and the Andamans. Widespread in the Pacific and Southeast Asia, extending to French Polynesia in the southeast and Hawaii in the northeast.

Holothuria flavomaculata Semper, 1868

COMMON NAMES: Red snakefish.

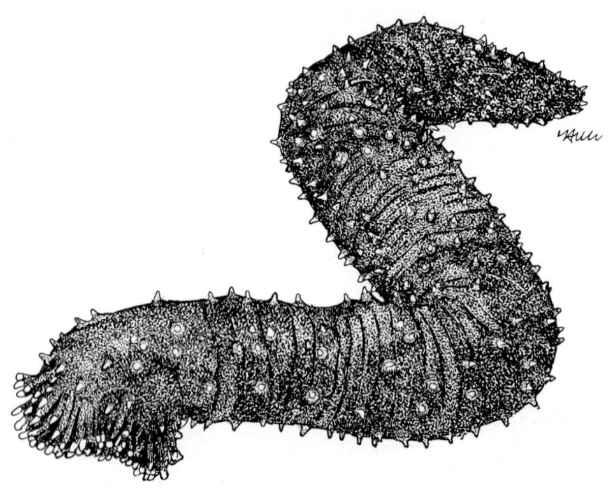

DIAGNOSTIC FEATURES: Dark grey, bluish-black, brownish-red to black over entire body, with characteristic pinkish, orangy or reddish tips to the numerous, large papillae over the body and yellowish tentacles. A relatively large, elongated sea cucumber. Some diagnostic characters in the literature are conflicting. Papillae are numerous on the lateral margins and around the mouth. Podia are more numerous near the posterior end. Mouth is ventral, with 20 to 31 greyish or yellowish tentacles with lighter terminal discs. Anus is terminal, encircled with 5 groups of papillae. Cuvierian tubules absent.

Ossicles: Tentacles with straight or curved rods, 95–355 µm long. Dorsal and ventral body wall with similar tables and rods. Tables without disc, spire ending in a Maltese cross. Rods spiny and massive, 85–105 µm long. Podia with tables similar to those of the body wall and, in addition, rods with perforated extremities, 160–200 µm long, and perforated plates, 130–210 µm long.

Processed appearance: Similar to *Holothuria coluber*, dried *H. flavomaculata* are elongated and irregular in shape, and clearly tapered at the anterior end. Brown body covered with lighter-coloured bumps. Small cut across mouth and/or in the body middle. Common size probably about 20 cm.

Remarks: This species may resemble *H. coluber*, especially in the Indian Ocean, and has been mistaken for that species in some reports.

Size: Maximum length 60 cm; average length 35 cm.

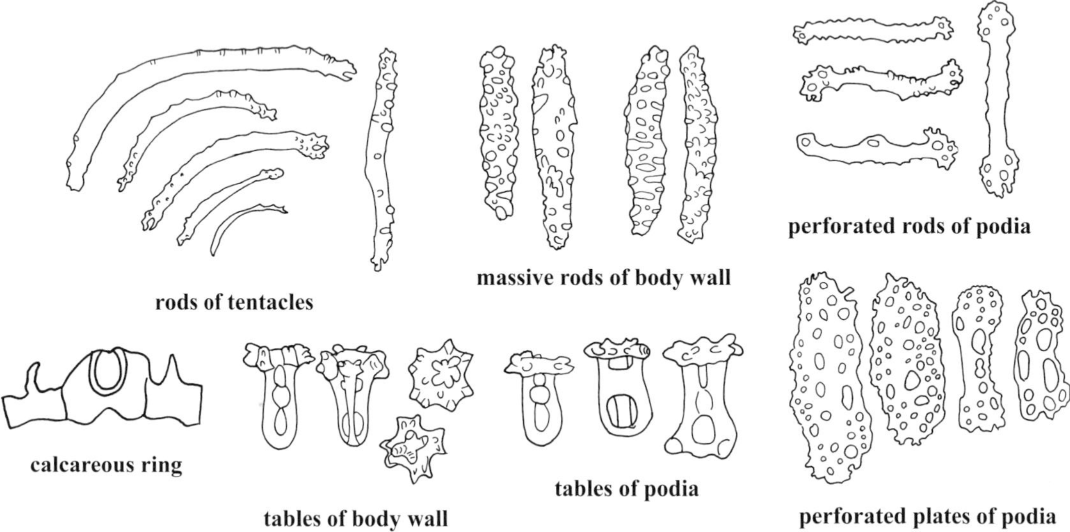

rods of tentacles

massive rods of body wall

perforated rods of podia

calcareous ring

tables of body wall

tables of podia

perforated plates of podia

(after Massin, 1999)

HABITAT AND BIOLOGY: Generally inhabits areas with mud, sand or coral rubble. Lives in waters from 1 to 40 m deep. It can be found with its posterior end hidden under coral rocks or crevices, but also may feed in the open. Populations generally at low density and often few individuals are recorded at each locality.
Its reproductive biology is unknown.

LIVE (photo by: S.W. Purcell)

EXPLOITATION:

Fisheries: In the western central Pacific, this species is commercially harvested in Palau, the Federated States of Micronesia, Solomon Islands and Vanuatu. It is probably also harvested artisanally in certain localities in the Indian Ocean and Southeast Asia.

Regulations: There are few regulations pertaining to the harvesting of this species.

Human consumption: Mostly, the reconstituted body wall (bêche-de-mer) is consumed by Asians.

Main market and value: This species is low value, similar to *H. coluber*. In Fiji, fishers occasionally collect it and sell it to processors for about USD2 kg^{-1} fresh gutted.

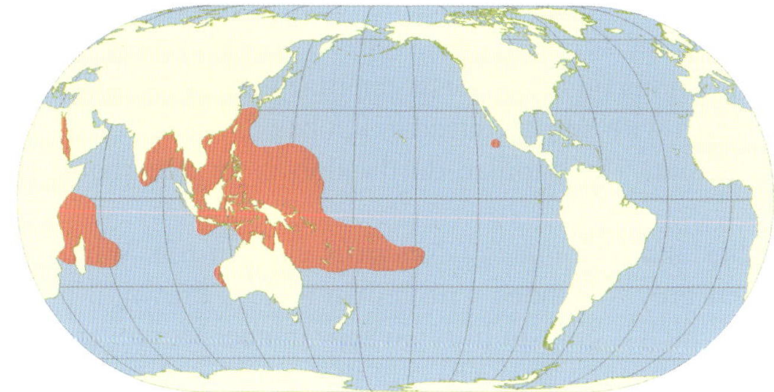

GEOGRAPHICAL DISTRIBUTION:
The Indian Ocean and western central Pacific. Reported from Madagascar, Mascarene Islands, the Red Sea, Sri Lanka, Indonesia, China, the Philippines, Australia, Palau, Guam, Micronesia (Federated States of), New Caledonia to French Polynesia and Clipperton Island.

Holothuria fuscocinerea Jaeger, 1833

COMMON NAME: Labuyo (Philippines).

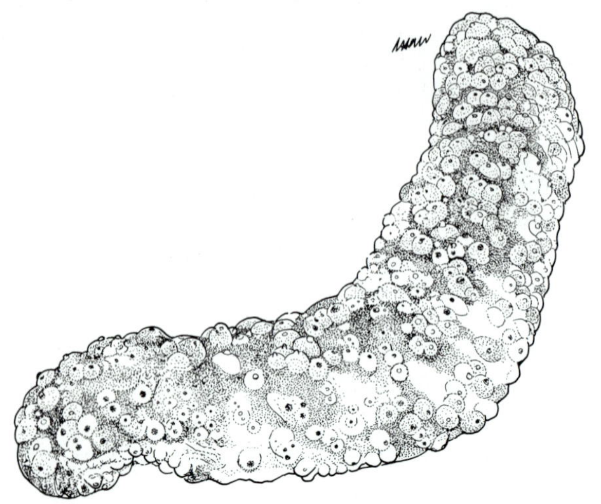

DIAGNOSTIC FEATURES: This species is greyish-brown or greyish-green dorsally and beige to brown ventrally. It may have brown blotches dorsally. It is a medium-to-large (>20 cm long), cylindrical species with a soft tegument. The brown papillae on the dorsal surface are wide at the base and narrow tipped. The podia of the ventral surface are sparse and small, but these are more numerous at the lateral flanks of the ventral surface. The ventral mouth has 20 large tentacles. Anus is dorsal and surrounded by a dark purple ring. It possesses very thick, numerous Cuvierian tubules, which are readily ejected.

Ossicles: Tentacles with curved rods, 50–400 µm long slightly rugose at extremites. Dorsal and ventral body wall with rather poorly developed tables and buttons. Table discs roundish and smooth, 25–40 µm across, perforated by 4 central and few peripheral holes, low spire ending in an ill-formed crown. Buttons, 25–40 µm long, smooth, irregular, with 1–3 pairs of holes. Ventral podia with irregular perforated rods, up to 235 µm long, large perforated plates, 100–155 µm long, buttons up to 70 µm long, and tables with spire reduced to knobs on disc. Dorsal papillae with rods, up to 300 µm long, perforated distally and some large tables with spire reduced to knobs.

Processed appearance: Light brown in colour. The papillae on the dorsal surface should be evident as bumps in dried specimens.

Remarks: The distinction between this species and *Holothuria pervicax* is clear in life. While the former can be variable in appearance from light grey to banded with brown to dark brown, it never has pinkish bases to its papillae as *H. pervicax*, which is quite uniform in appearance. Moreover, only *H. fuscocinerea* has a dark brown ring about its anus and near the base of the ventral podia.

Size: Maximum length 30 cm; average length is about 20 cm.

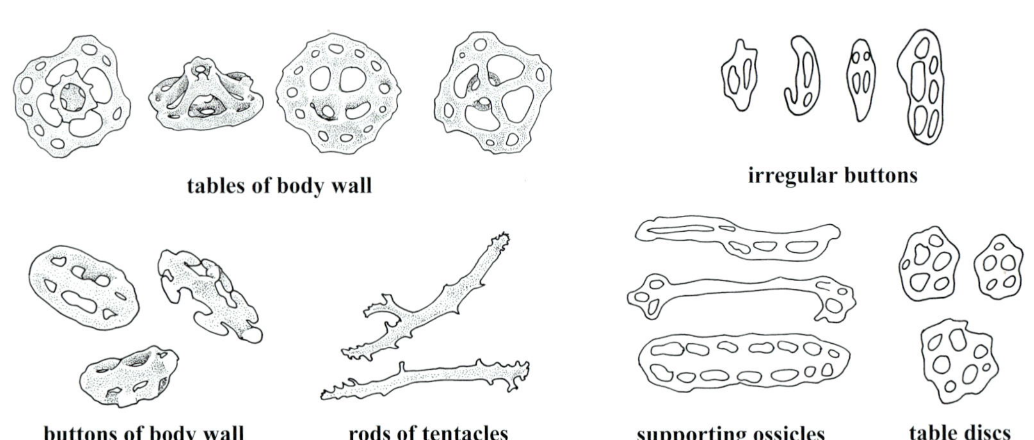

tables of body wall

irregular buttons

buttons of body wall   rods of tentacles

supporting ossicles   table discs

(source: Solís-Marín *et al.*, 2009)

(after Reyes-Leonardo, 1984)

HABITAT AND BIOLOGY: Prefers habitats between 0 and 30 m depth. Found in lagoonal habitats, reef flats and on outer reef slopes of barrier reefs. In Kenya, it is a nocturnal species.
Its reproductive biology is unknown.

EXPLOITATION:

Fisheries: This species does not have a commercial value in the western central Pacific; however, it is of commercial importance in China, Malaysia and the Philippines. In China, it is of low commercial importance. It is exploited in a multispecies fishery in Sri Lanka.

Regulations: There are few regulations pertaining to the harvesting of this species.

Human consumption: Not avalable.

Main market and value: Probably low value. It has been reported to have been traded at up to USD3 kg^{-1} dried in the Philippines.

LIVE (photo by: G. Edgar)

PROCESSED (photo by: J. Akamine)

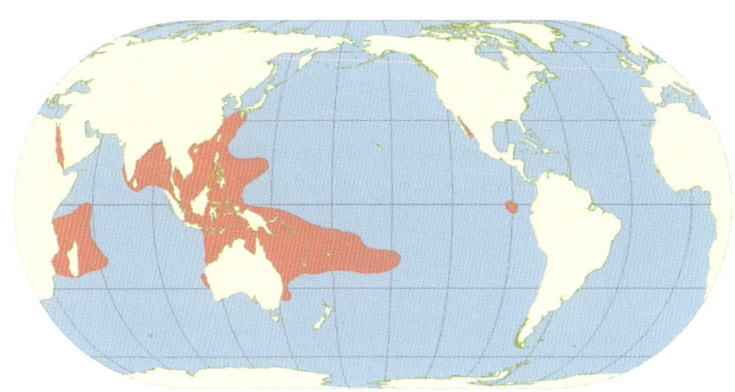

GEOGRAPHICAL DISTRIBUTION:
It can be found in the the Indian Ocean, Red Sea, western central Pacific and Asia. Also distributed in Celebes and Amboina, Sri Lanka, Bay of Bengal, East Indies, northern Australia, the Philippines, China, southern Japan, Guam and South Pacific Islands. Reported also from Galapagos Islands and Gulf of California.

Holothuria fuscogilva Cherbonnier, 1980

COMMON NAMES: White teatfish (**FAO**), White mammyfish (India), Holothurie blanche à mamelles (**FAO**), Kal attai (India), Bawny white (Egypt), Pauni myeupe (Zanzibar, Tanzania), Benono (Madagascar), Le Tété blanc (New Caledonia), Susuan (Philippines), Huhuvalu hinehina (Tonga), Temaimamma (Kiribati), Sucuwalu (Fiji).

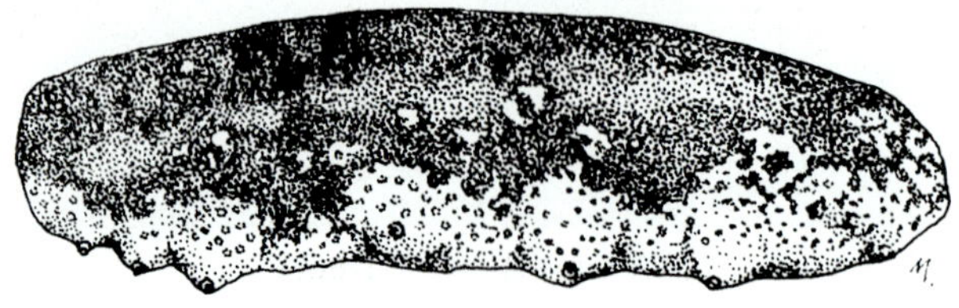

DIAGNOSTIC FEATURES: Colour variable, from completely dark brown, to dark grey with whitish spots, or whitish or beige with dark brown blotches. In the Western Indian Ocean, it tends to be reddish-brown dorsally and white ventrally and the anus is yellow. Ventral surface is greyish to brown. Body is suboval, strongly flattened ventrally, stout and quite firm with a thick body wall, and presents characteristic large lateral protrusions ('teats') at the ventral margins. Podia on the dorsal surface are sparse and small, but these are numerous on the ventral surface. The tegument is usually covered by fine sand. Mouth is ventral with 20 stout grey tentacles. Anus surrounded by inconspicuous teeth. No Cuvierian tubules. Juveniles are yellowish-green or yellow, with black blotches.

Ossicles: Tentacles with stout rods, up to 700 µm long, rugose distally. Dorsal body wall with tables and ellipsoid buttons. Table disc roundish and undulating, 65–100 µm across, perforated by 10–15 holes, low spire ending in a stout crown of spines that can have more than one layer in the largest tables. Ellipsoid buttons irregular, some 65 µm long. Ventral body wall with similar tables and ellipsoid buttons as those dorsally, and, in addition, slightly knobbed buttons, 60–80 µm long. Ventral and dorsal podia with large perforated plates.

Processed appearance: Flat and stout shape with obvious teats along sides. Surface smooth to slightly wrinkled and powdery. Entire body different shades of grey-brown. One single cut dorsally but not completely to the mouth or anus. Common size 18–24 cm.

Remarks: Uthicke, O'Hara and Byrne (2004) give this species a wide Indo-Pacific distribution and separate it from *Holothuria whitmaei*. The ventral surface is light grey or brownish, whereas it is dark grey in *H. whitmaei*. The lateral 'teats' may appear longer and thinner than in most individuals of *H. whitmaei*.

Size: Maximum length about 57 cm. Average fresh weight from 2 400 g (Madagascar, India and Papua New Guinea) to 3 000 g (Egypt); average fresh length from 40 cm (India and Madagascar), 42 cm (Papua New Guinea) to 60 cm (Egypt). In New Caledonia, average live weight about 2 440 g and average live length about 28 cm.

HABITAT AND BIOLOGY: Commonly inhabits outer barrier reef slopes, reef passes

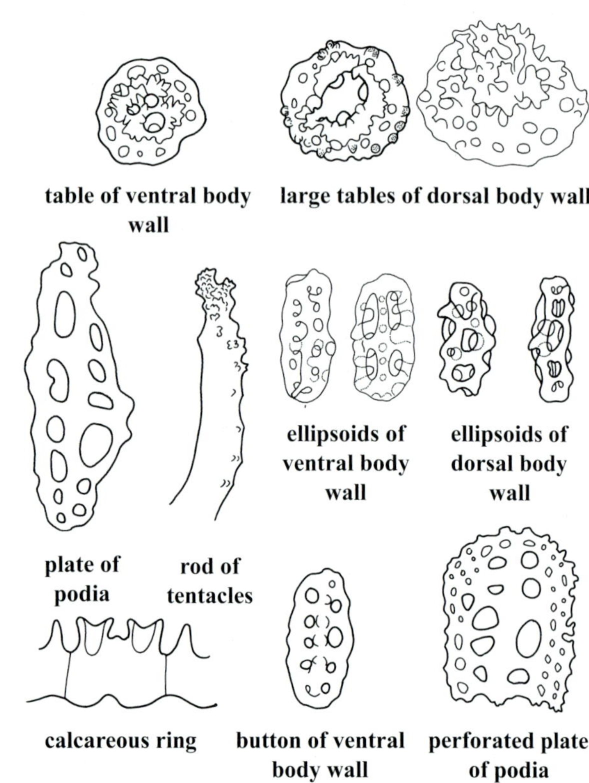

table of ventral body wall

large tables of dorsal body wall

ellipsoids of ventral body wall

ellipsoids of dorsal body wall

plate of podia

rod of tentacles

calcareous ring

button of ventral body wall

perforated plate of podia

(after Cherbonnier, 1980)

and sandy areas in semi-sheltered reef habitats in 10 to 50 m water depth. Can also be found in seagrass beds (Papua New Guinea and India) between 0 and 40 m. In Fiji, this species recruits in shallow seagrass beds and then moves to deeper zones.

It attains size-at-maturity at 1 100 g. In New Caledonia, this species reproduces between November and January, while in Solomon Islands between August and October.

LIVE (photo by: S.W. Purcell)

EXPLOITATION:

Fisheries: Harvested in artisanal (e.g. the Philippines, Tonga, and Madagascar), semi-industrial (New Caledonia) and industrial fisheries (Australia, Egypt) throughout its range in the Indo-Pacific, and is among the most valuable species. Harvested by hand collecting, free diving and lead-bombs and by SCUBA diving (Madagascar) and hookah

PROCESSED (photo by: S.W. Purcell)

(Australia). In many fisheries *H. fuscogilva* has been overexploited. In the Africa and Indian Ocean region, it is fished in the Comoros, Mozambique, Kenya, Madagascar and Seychelles. In Seychelles, it is considered fully exploited.

Regulations: Before a moratorium in Papua New Guinea, regulations included a minimum size limit (35 cm live and 15 cm dry). On the Great Barrier Reef, Australia, there is an overall TAC of 89 tonnes y^{-1}, which is reviewed periodically. In other fisheries in Australia, a size limit of 32 cm is imposed. In New Caledonia, the minimum size limit is 35 cm for live animals and 16 cm dried, and harvesting using compressed air is prohibited. In Maldives, there is a ban on the use of SCUBA to protect the stocks of this species.

Human consumption: Mostly, the reconstituted body wall (bêche-de-mer) is consumed by Asians and is highly regarded.

Main market and value: It is a high-value species. In Papua New Guinea, it was previously sold at USD17–33 kg^{-1} dried. It has been traded recently at USD42–88 kg^{-1} dried in the Philippines. In New Caledonia, this species is exported for USD40–80 kg^{-1} dried and fishers may receive USD7 kg^{-1} wet weight. In Fiji, fishers receive USD30–55 per piece fresh. Prices in Hong Kong China SAR retail markets ranged from USD128 to 274 kg^{-1}. Prices in Guangzhou wholesale markets ranged from USD25 to 165 kg^{-1} dried.

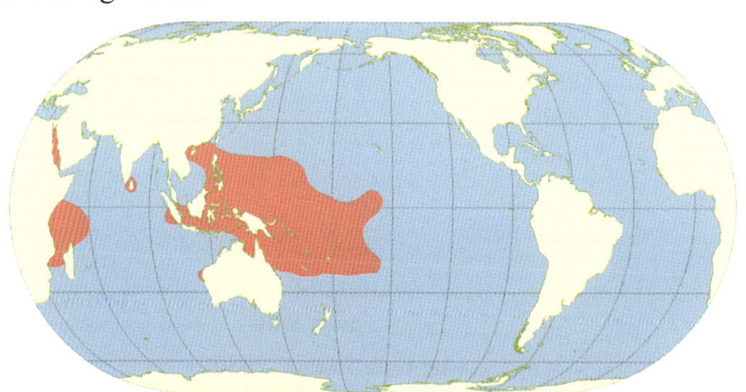

GEOGRAPHICAL DISTRIBUTION:

From Madagascar and the Red Sea in the west, across to Easter Island in the east and from southern China to south to Lord Howe Island. Occurs throughout much of the western central Pacific as far east as French Polynesia.

Holothuria fuscopunctata Jaeger, 1833

COMMON NAMES: Elephant trunkfish (**FAO**), Holothurie trompe d'éléphant (**FAO**), Betaretry (Madagascar), Barangu mwamba (Zanzibar, Tanzania), Sapatos (Philippines), Ngoma (Kenya), Kunyi (Indonesia), L'éléphant (New Caledonia), Terebanti (Kiribati), Elefanite (Tonga), Dairo-ni-cakao (Fiji).

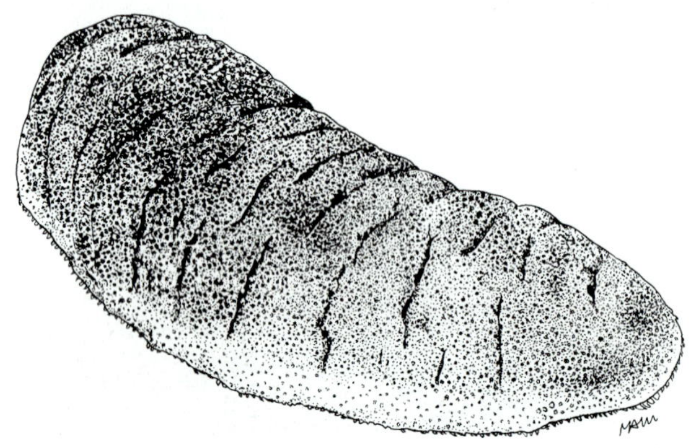

DIAGNOSTIC FEATURES: Coloration varies a little from golden to light brown or creamy dorsally with numerous brown spots (around papillae), shading to whitish ventrally. This species has characteristic deep, brown wrinkles dorsally (like part of an elephant's trunk). Body is suboval, arched dorsally and strongly flattened ventrally. A large species with a stout body and thick body wall. The body is often covered by fine sediment. Mouth ventral with 20 stout, brown, tentacles. Anus is large and black, has no teeth, and is surrounded by five groups of papillae. No Cuvierian tubules.

Ossicles: Tentacles with straight rods, 30–150 µm long, slightly spiny. Dorsal and ventral body wall with numerous tables and ellipsoid buttons, with ventrally also some smooth and knobbed buttons. Tables have small discs, 35–55 µm across, with irregular and spiny rim, perforated by 4 central and few peripheral holes, and a low spire that ends in a spiny crown. Ellipsoid buttons perforated by 4–6 pairs of holes, on average 75 µm long. Dorsal and ventral podia with spiny plates which can take the form of irregular branching rods.

Processed appearance: Processed animals are relatively elongate, arched dorsally, flattened ventrally. Light brown to beige dorsally with deep grooves. Ventral surface is smoother. Tiny black spots are noticeable over whole body. Small cut across mouth or one single long cut ventrally. Common dried size 20–25 cm.

Remarks: Cherbonnier (1980) considered *Holothuria axiologa* Clark, 1921 a junior synonym. The body fluid is bright yellow and stains, making this species undesirable to harvest and gut.

Size: Maximum length about 70 cm; average length about 48 cm. Average adult weight 3 kg; maximum 5.5 kg.

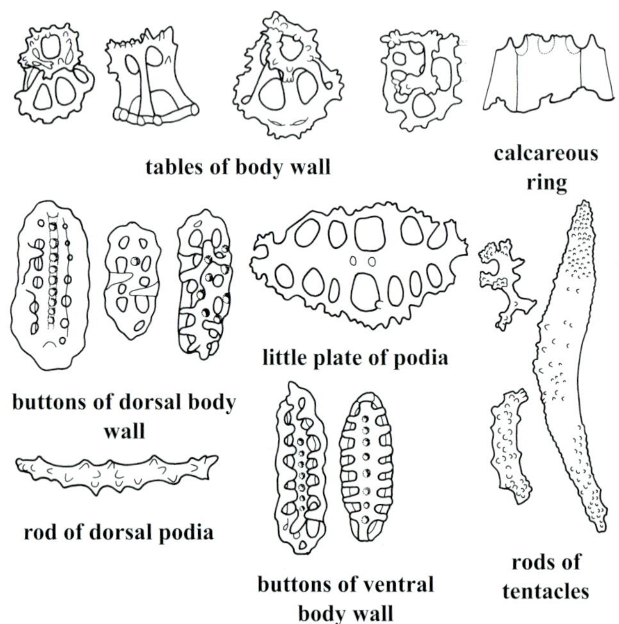

tables of body wall

calcareous ring

buttons of dorsal body wall

little plate of podia

rod of dorsal podia

buttons of ventral body wall

rods of tentacles

(after Cherbonnier, 1980)

HABITAT AND BIOLOGY:

H. fuscopunctata lives in shallow waters, generally from 3 to 25 m depth. Inhabits reef slopes, lagoons and seagrass beds over sandy bottoms. Generally found on coarse sand or coral rubble.

It attains size-at-maturity at 1 200 g. On the Great Barrier Reef (Australia), this species reproduces annually in December, while in New Caledonia from December to February.

LIVE (photo by: S.W. Purcell)

EXPLOITATION:

Fisheries: This species is mostly fished artisanally. Harvested mostly by hand collecting by free-diving. Also collected using lead-bombs (e.g. Papua New Guinea) and SCUBA diving (Madagascar). In the western Pacific region, it is commercially exploited in most localities, east to Tonga. Previously harvested in Papua New Guinea,

PROCESSED (photo by: S.W. Purcell)

Solomon Islands and Vanuatu before a general moratorium. Harvesting is minimal in New Caledonia, Coral Sea and Torres Strait (Australia). In Tuvalu, it comprises 8% of the total catches. In Seychelles, this species is currently underexploited.

Regulations: Before a fishery moratorium in Papua New Guinea, fishing for this species was regulated by minimum landing size limits (45 cm live; 15 cm dry) and other regulations. In Torres Strait (Australia), there is a size limit of 24 cm live.

Human consumption: Mostly the reconstituted body wall (bêche-de-mer) is consumed by Asians. Despite its thick body wall, the market price is relatively low because the flesh is bitter and numbs the mouth.

Main market and value: China. It has been traded recently at about USD8 kg^{-1} dried in the Philippines. In Papua New Guinea it was previously sold at USD2.7 kg^{-1} dried. In Fiji, fishers receive USD0.8–1.7 per piece fresh. Prices in Guangzhou wholesale markets ranged from USD11 to 19 kg^{-1} dried.

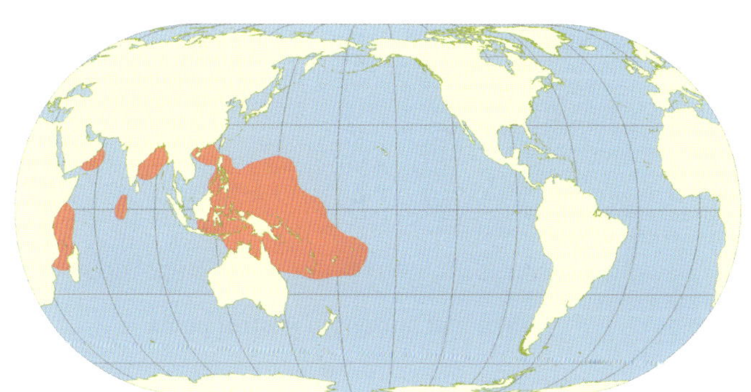

GEOGRAPHICAL DISTRIBUTION:

It can be found in the western central Pacific, Asia and the Africa and Indian Ocean. *H. fuscopunctata* occurs at least as far east as Tonga.

COMMON NAMES: Tiger Tail, Mani-mani (Peanut-like), Bat-tuli (Philippines).

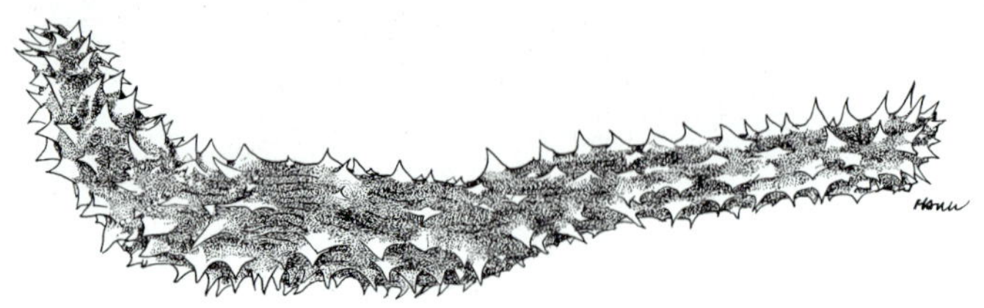

DIAGNOSTIC FEATURES:

Dorsal coloration is tan to reddish-brown with large, conical, yellow-to-whitish papillae. Ventral surface is beige-brown with white-yellow spots corresponding to large podia that are distributed in three to four rows. Body is cylindrical. Body wall smooth and just a few millimetres thick. Mouth is ventral and surrounded by 20 very short tentacles. Anus is terminal. Numerous Cuvierian tubules but never ejected.

Ossicles: Tentacles with slender rods, 45–145 μm long. Dorsal and ventral body wall with similar tables and rods, although there are fewer tables in the ventral body wall. Table discs, 50–70 μm across, smooth, circular to quadrangular in outline, perforated by four central holes and 9–13 peripheral ones; the short spire terminates in a narrow crown of spines. Buttons are 70–100 μm long, irregular, with smooth rim, an 3–6 pairs of holes. Ventral podia with buttons similar to those in the body wall and with perforated plates, up to 160 μm long and 75 μm wide. Dorsal papillae with buttons, up to 125 μm long and rods, up to 200 μm long.

Processed appearance: Chestnut colour more defined after boiling.

Size: Maximum length about 25 cm.

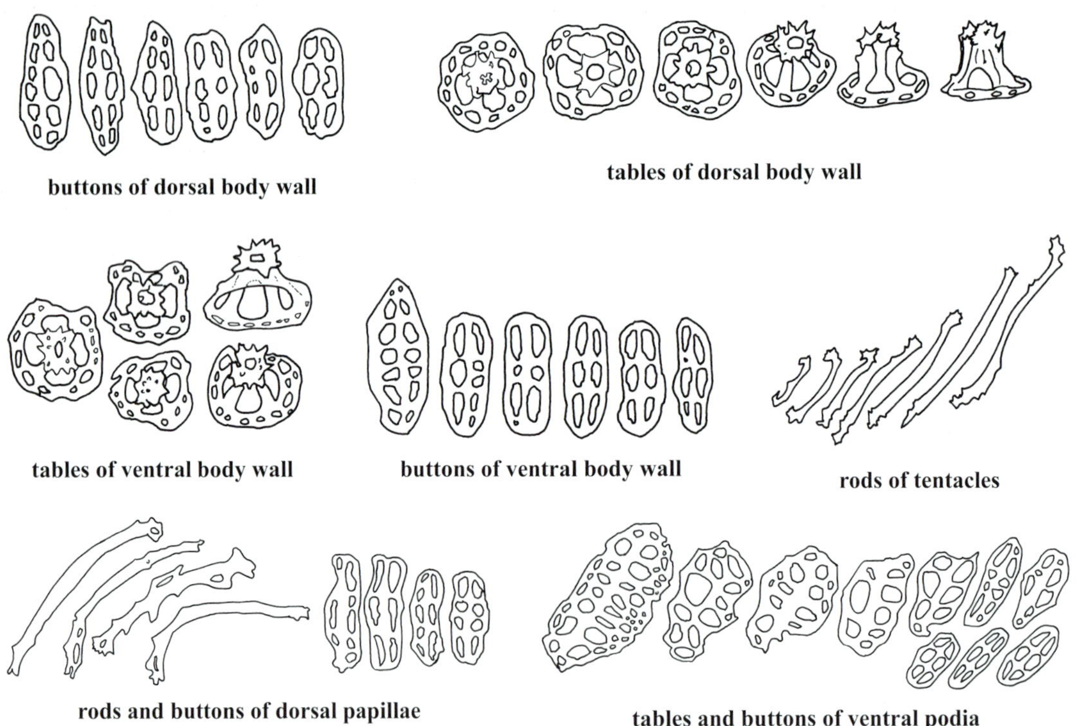

buttons of dorsal body wall

tables of dorsal body wall

tables of ventral body wall

buttons of ventral body wall

rods of tentacles

rods and buttons of dorsal papillae

tables and buttons of ventral podia

(after Samyn and Massin, 2003)

HABITAT AND BIOLOGY: Found under coral debris and under coral slabs within seagrass beds and on sandflats and sandy reef flats. In the Philippines, this species is found mostly in depths between 1 and 2 m. In the Comoros, it can be found mainly on rocks over coarse sandy bottoms between 0 and 20 m. Little is known of its reproductive biology.

LIVE (photo by: S.W. Purcell)

EXPLOITATION:

Fisheries: This species is mostly fished artisanally by hand collecting by gleaners on shallow reef flats or by free diving. In the western central Pacific it is not fished commercially but occasionally it is harvested for subsistence, e.g. in Samoa and Cook Islands. *Holothuria hilla* is fished commercially in the Philippines and Indonesia; in the latter it is part of a multispecies fishery, where it is used as a filler to top up weights sales. It is commercially harvested in Madagascar. In some Pacific island countries, it is collected for the aquarium trade.

Regulations: Where it is fished, there are few, or no, regulations pertaining to the harvesting of this species.

Human consumption: Mostly the reconstituted body wall (bêche-de-mer) is consumed by Asians. Its fermented intestine (konowata) may be eaten as a delicacy or as a protein component in traditional diets. In the Philippines, it is consumed by Muslims during the Ramadan season.

Main market and value: Poorly known, but it is a low-value species. It has been traded recently at about USD3 kg^{-1} dried in the Philippines.

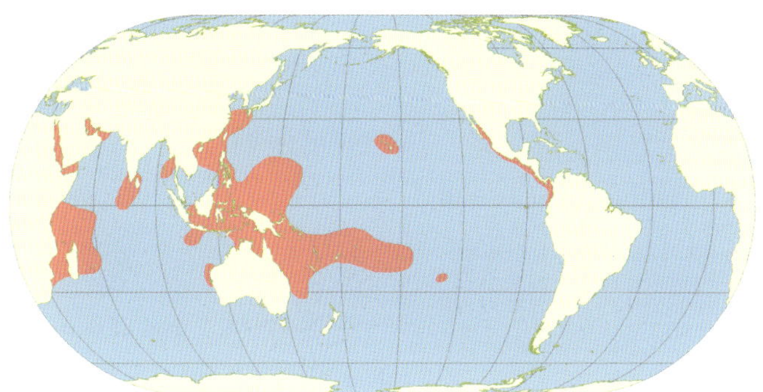

GEOGRAPHICAL DISTRIBUTION: Most countries of the western central Pacific to Pitcairn Islands and reported from eastern Central America. Found in Southeast Asia, east Africa and Indian Ocean.

Holothuria impatiens (Forsskål, 1775)

COMMON NAMES: Brown spotted sea cucumber, Impatient sea cucumber, Bottleneck sea cucumber (**FAO**), Holothurie bouteille (**FAO**), Holoturia cuello de botella (**FAO**), Sunlot (Philippines).

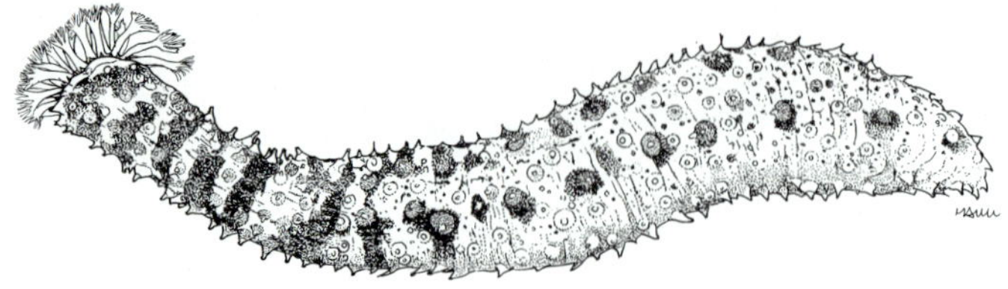

DIAGNOSTIC FEATURES: Coloration is light brown with 5 or more dark brown transverse bands on the dorsal surface which become spots posteriorly. Ventral surface is beige. The body of this species has been described as bottle shaped and is rough to the touch. Podia are relatively sparse. The ventral mouth contains 20 tentacles. Long, white, thick Cuvierian tubules are present.

Ossicles: Tentacles with straight and curved rods, 75–350 µm long, spiny at the extremities. Ventral and dorsal body wall with similar tables and buttons. Tables with round, smooth disc, 80–90 µm across, perforated by 4 large and 4–8 peripheral holes; short spire ending in a spiny crown. Buttons are 60–100 µm long, with a smooth rim and 3–4 pairs of holes, and sometimes with median line. Ventral and dorsal podia with tables and buttons similar to those of body wall and rods, 175–270 µm long with median and distal swellings and perforations.

Processed appearance: Dried individuals are small (about 5–7 cm in length) and are generally mixed with other small low-value species.

Remarks: *Holothuria impatiens* is a species complex. Ongoing taxonomic investigation is untangling it at present.

Size: Maximum length about 26 cm. Average fresh weight 50 g; average fresh length probably <20 cm.

HABITAT AND BIOLOGY: *H. impatiens* is a cryptic species. It lives in shallow coral reef habitats. In the Comoros and Mascarene Islands, it can be found under rocks in shallow waters between 0 and 2 m; however, it can be observed up to 30 m depth. Its reproductive biology is unknown.

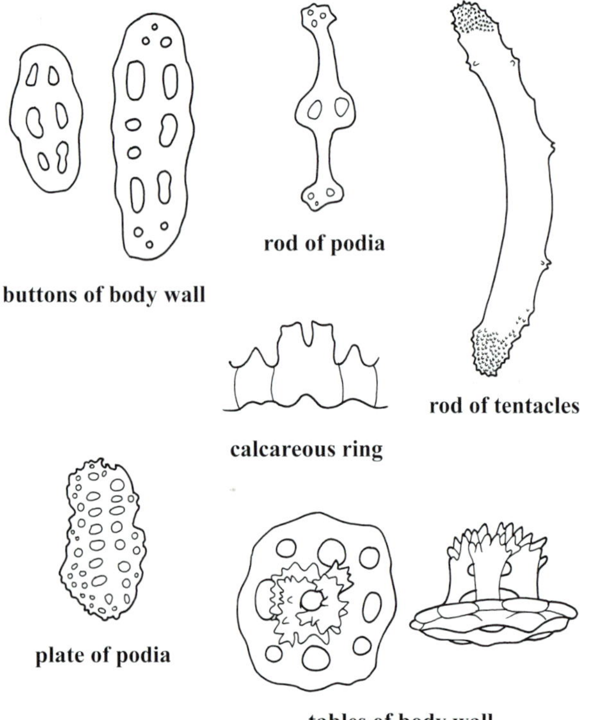

buttons of body wall

rod of podia

calcareous ring

rod of tentacles

plate of podia

tables of body wall

(after Cherbonnier, 1980)

EXPLOITATION:

Fisheries: This species is harvested in artisanal fisheries, by hand collection in shallow waters. It is harvested by hookah diving in south Viet Nam. In the western central Pacific it does not have a commercial value, so it is unexploited in that region. It has commercial importance in China and Indonesia and Mexico. In Madagascar, there is limited harvesting.

Regulations: None.

Human consumption: Mostly, the reconstituted body wall (bêche-de-mer) is consumed by Asians.

Main market and value: This is a low-value species, principally exported to Chinese markets. It has been traded recently at about USD2 kg^{-1} dried in the Philippines.

LIVE (photo by: S. Ribes)

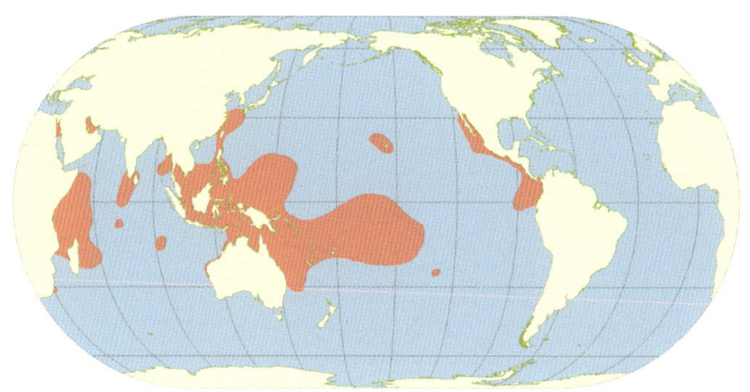

GEOGRAPHICAL DISTRIBUTION:

It can be found from East Africa and the Indian Ocean to the western central Pacific including Hawaii, in the Pitcairn Islands group, and including much of the Pacific coast of Central America.

COMMON NAME: Sea cucumber.

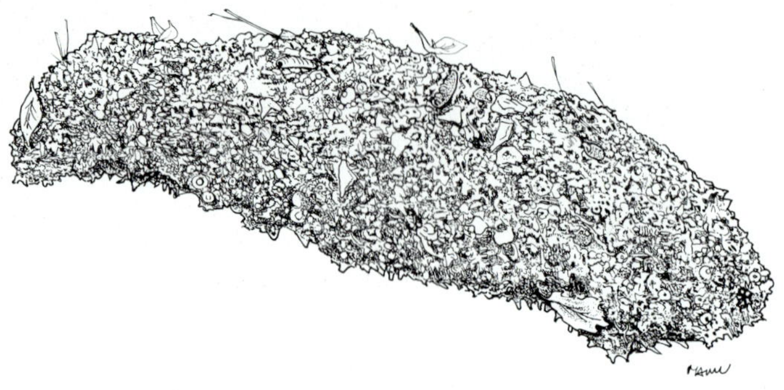

DIAGNOSTIC FEATURES:
Holothuria kefersteini is a medium-sized species. Coloration is reddish-brown to greyish with dark-tipped papillae. Dorsal surface generally has four to six small, wart-like bumps and a number of small papillae. Mouth ventral with 20 brown to black tentacles. Body often covered with sand and detached pieces of algae.

Ossicles: In the body wall there is an external layer of tables with small (40–50 µm), or completely reduced, disc that often has some prominent marginal spines, and the well-developed spire ends in a Maltese cross. The inner layer of the body wall holds small, c.a. 50 µm long, perforated plates, mostly with 2–4 large central holes and some small terminal ones. Ventral podia with rods, 80 µm long, with enlarged, perforated endings. Dorsal papillae with straight rods with perforated ends.

Processed appearance: Dark brown to black in colour. Dorsal surface highly textured and rough, with rugosities. Ends relatively blunt.

Size: Maximum size up to 20 cm.

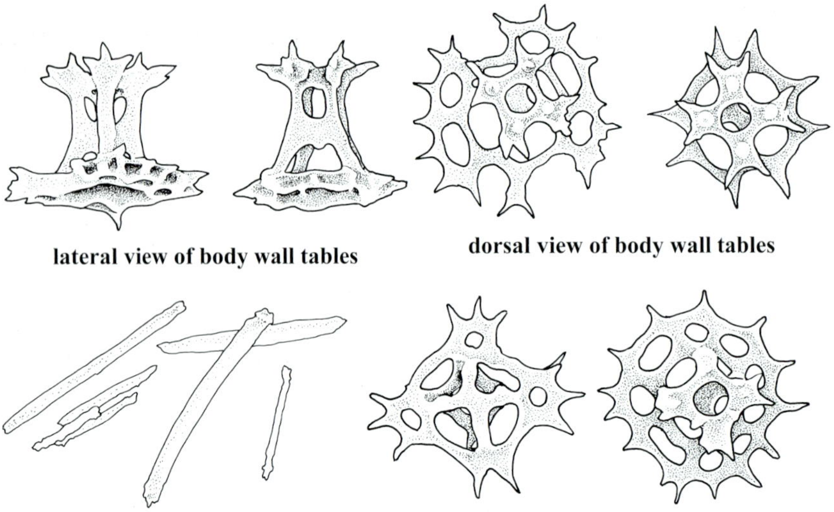

lateral view of body wall tables   dorsal view of body wall tables

rods of podia   perforated plates of body wall

(source: Solís-Marín *et al.*, 2009)

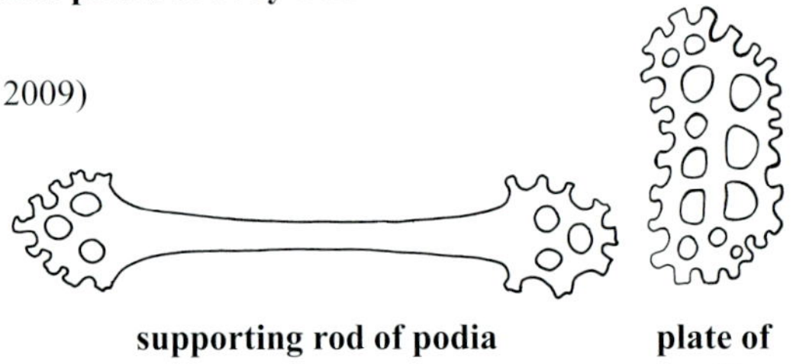

supporting rod of podia   plate of podia

(source: Deichmann, 1958)

HABITAT AND BIOLOGY: In El Salvador, this species appears to prefer rocky shores. In the Galapagos Islands, it can be found in the intertidal and subtidal zone, generally exposed on coral sand bottoms. It is often the most common species on sandy bottoms. It can be found to a maximum depth of 18 m. Reproductive biology is unknown.

LIVE (photo by: G. Edgar)

EXPLOITATION:

Fisheries: This species is fished artisanally. It is exploited illegally in the Galapagos Islands, El Salvador and Mexico, where it is reported to be severely over-exploited. In the Galapagos Islands, it is harvested by hand collecting using hookah diving. In El Salvador, it is part of a multispecies fishery, which probably includes *Isostichopus fuscus*.

PROCESSED (photo by: S.W. Purcell)

Regulations: All species of holothurians in El Salvador are listed under the Endangered Species list of the Ministry of Environment.

Human consumption: Mostly, the reconstituted body wall (bêche-de-mer) is consumed by Asians.

Main market and value: Retail prices in Hong Kong China SAR were up to USD126 kg^{-1} dried.

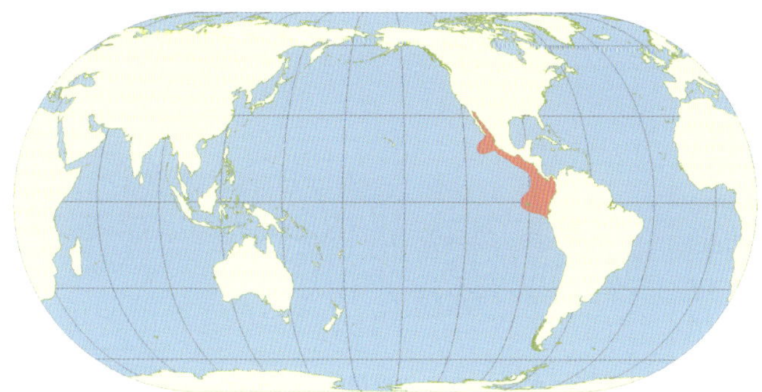

GEOGRAPHICAL DISTRIBUTION: Gulf of Baja California, Baja, Central America, Colombia, Ecuador and Peru and the oceanic islands of Revillagigedos, Galapagos Islands, Cocos Island and Malpelo.

COMMON NAMES: Golden sandfish, Zanga mena, Chinois, Matafao (Madagascar), Barangu (Zanzibar, Tanzania), Curtido-bato (Philippines), Bangkuli (Indonesia), Le mouton (New Caledonia), Nga ito (Tonga).

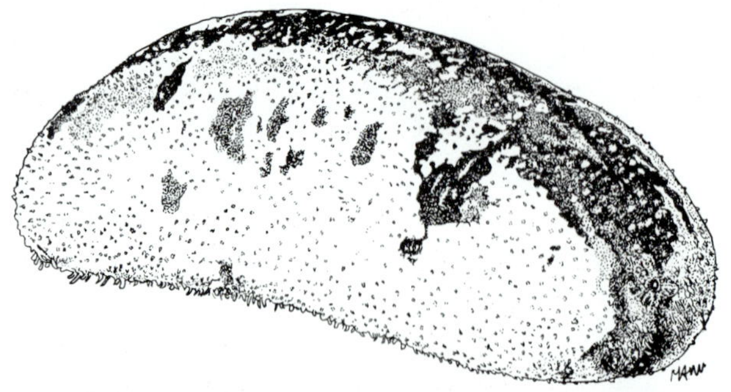

DIAGNOSTIC FEATURES: Coloration highly variable: from dark greyish black, to beige with black blotches and spots, or beige without black spots. Ventral surface is whitish, or grey in the black variants. Body is relatively stout, highly arched dorsally but flattened ventrally. Ends of the body are rounded. Body is often covered by fine sediments and mucus. Dorsal surface without deep wrinkles, but with relatively large, black papillae (3 mm long). Brown podia are moderately abundant ventrally. The ventral mouth has 20 short grey tentacles. The anus is terminal, surrounded by five groups of papillae and without teeth. Cuvierian tubules absent.

Ossicles: Tentacles with rods, 60–650 µm long, slightly curved with spiny extremities. Dorsal body wall: tables with disc 50–110 µm across, that are spiny, quadrangular, and perforated by one central hole and 4–10 peripheral holes (large discs with circles of peripheral holes); spire ending in a spiny crown; buttons 40–60 µm long, knobbed, and have 3–4 pairs of small holes. Ventral podia with tables and buttons as in body wall (but some are smooth) and with perforated rods, 115–265 µm long, and perforated plates, 85–280 µm long. Dorsal podia with buttons, tables and rods. Buttons are smooth or nodulous, with 3–4 pairs of holes, while the perforated plates are 160–200 µm long and have two rows of holes.

Processed appearance: Entire body golden-brown colour, small cut ventrally. Relatively elongate with round ends, slightly arched dorsally. Common dried size 13 cm.

Remarks: Previously known as *Holothuria scabra* var. *versicolor* (Conand, 1989). There are apparently no marked differences in coloration between Indian Ocean and Pacific populations.

Size: Maximum length 46 cm; average length 30 cm. Average fresh weight: 1 100 g (Madagascar), 1 355 g (Australia), 1 400 g (New Caledonia); average fresh length: 30 cm (Australia, New Caledonia and Madagascar).

HABITAT AND BIOLOGY: Lives in many shallow water habitats up to 20 m depth (often deeper than *H. scabra*). Individuals can become larger than *H. scabra*. Occurs on inner reef flats and coastal lagoons or coastal seagrass beds or sandflats between 0 and 30 m depth. It buries in sand or muddy-sand during parts of the day. In Madagascar, it can be found in the inner slopes and seagrass beds, with highest densities in the latter. It attains size-at-maturity at 480 g (Madagascar). Sexual reproduction occurs annually; in late spring (Australia) or in the dry season (Madagascar).

EXPLOITATION:
Fisheries: Exploitation may be industrial (Australia) or artisanal (e.g. New Caledonia,

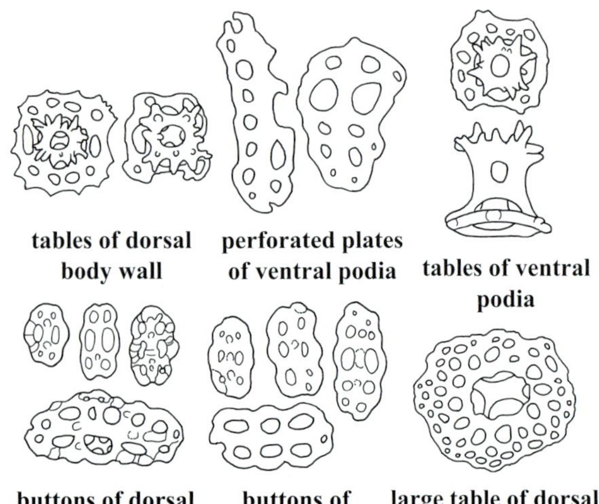

(after Massin *et al.*, 2009)

Madagascar) or in small quantities for subsistence (e.g. Fiji, Tonga). In different localities, this species is collected using free diving, SCUBA diving or using hookah. It is exploited throughout most of its distributional range. Presently banned by moratoria in Papua New Guinea, Solomon Islands and Vanuatu. Exploited in New Caledonia, Fiji, Tonga, Torres Strait and the Great Barrier Reef (Australia), Indonesia, the Philippines, Madagascar and Seychelles.

Regulations: In Australia, there is a minimum size limit of 17 cm fresh, no-take zones, strict licensing, and this species is subject to a rotational harvest strategy. In Tonga, it is banned from export to safeguard stocks for local consumption. In New Caledonia, there is a minimum size limit of 30 cm for live animals and 11 cm dried.

Human consumption: Mostly, the reconstituted body wall (bêche-de-mer) is consumed by Asians and is highly regarded. In a small number of Pacific island nations (e.g. Fiji and Tonga), it is consumed locally and is cooked with coconut milk.

Main market and value: Asia. In Australia, it is sold at about USD8 kg^{-1} gutted weight and exported for >USD100 kg^{-1}. In New Caledonia, this species is the second-most valuable species; fishers may receive USD4–6 kg^{-1} wet (gutted) weight and it is exported for about USD70–80 kg^{-1} dried. In Fiji it is banned for export but fishers still sell it and receive about USD8 per piece fresh. Prices in Hong Kong China SAR retail markets ranged from USD242 to 787 kg^{-1} dried.

LIVE (blotchy variant) (photo by: S.W. Purcell)

LIVE (beige variant) (photo by: S.W. Purcell)

LIVE (black variant) (photo by: S.W. Purcell)

PROCESSED (photo by: B. Giraspi)

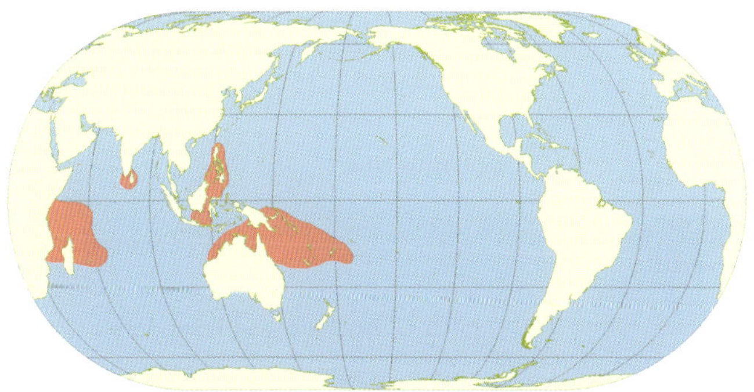

GEOGRAPHICAL DISTRIBUTION: Throughout the Indo-Pacific. Countries of known sightings include Madagascar, the Comoros, India, Kenya, Mayotte, Seychelles, Indonesia, the Philippines, Papua New Guinea, northern Australia, New Caledonia, Vanuatu, Fiji, and as far east as Tonga.

Holothuria leucospilota Brandt, 1835

COMMON NAMES: White threadsfish (**FAO**, Réunion), Snakefish (Viet Nam), Trépang à canaux blancs (**FAO**), Patola (Philippines), Zanga kida (Madagascar), Sumu (Tanzania), Kichupa (Zanzibar, Tanzania).

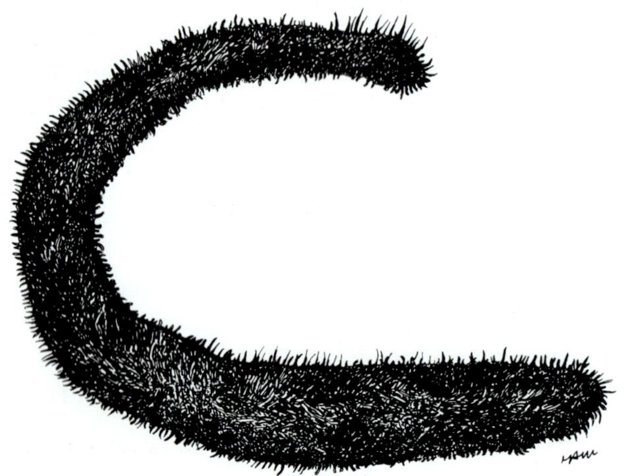

DIAGNOSTIC FEATURES: Body is entirely black, elongated, and somewhat broader in the posterior half. Body tapers moderately at both anterior and posterior ends. Long podia and papillae are randomly distributed on the dorsal surface. Ventral podia are numerous. The tegument is sometimes covered by fine sediment and mucus. Mouth is ventral with 20 large black tentacles. Anus is terminal and this species ejects thin, long Cuvierian tubules. Juveniles have a similar appearance to adults.

Ossicles: Tentacles devoid of ossicles. Dorsal and ventral body wall with similar tables and buttons. Tables with round to quandrangular discs, 40–70 µm across, perforated by 4 central holes and 4–12 peripheral ones, and the rims of discs are smooth to spiny; spire ending in a spiny crown. The irregular buttons are 40–70 µm long, with 2–5 pairs of irregular holes. Ventral podia with similar tables and buttons and with large perforated plates, 60–120 µm long. Dorsal podia with tables and buttons similar to those of body wall and rods, 50–190 µm long, variously perforated.

Processed appearance: This species may be traded mixed with other low-value species in the dried form.

Size: Maximum length about 50 cm; average length about 30 cm. Average fresh weight from 335 g (Viet Nam) to 400–900 g (Réunion); average fresh length from 23 cm (Viet Nam) to 35–50 cm (Réunion).

HABITAT AND BIOLOGY: Lives in shallow habitats up to 10 m depth. Found mostly on outer and inner reef flats, back reefs and shallow coastal lagoons. Commonly found in seagrass beds, sandy and muddy bottoms with rubble or coral reefs. *Holothuria leucospilota* is a very common species, with its distribution extending into warm-temperate zones. Densities may be up to 5 000 ind. ha^{-1}. In Madagascar, it can be found in the inner slope, seagrass beds, microatoll and detrital fringe with higher densities on the inner reef slopes.

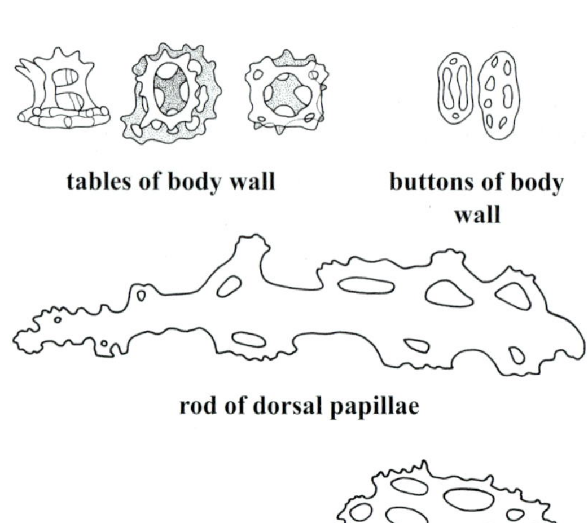

tables of body wall

buttons of body wall

rod of dorsal papillae

calcareous ring

plate of ventral podia

(after Cherbonnier and Féral, 1984)

This species attains size-at-maturity at 180 g and sexual reproduction occurs bi-annually, during the dry season. Smaller individuals may reproduce asexually by transverse fission. On the Great Barrier Reef (Australia), it reproduces sexually between November and March, while in the Northern Territory (Australia) in April. In the Cook Islands, this species reproduces from October to April. In Taiwan Province of China, it reproduces between June and September. In Réunion, it reproduces sexually twice a year; in February and in May.

LIVE (photo by: S.W. Purcell)

EXPLOITATION:

Fisheries: This species is harvested in artisanal fisheries at localities where low-value species are exploited. Harvested predominantly by hand collection at low tide and by free diving. In the southern Cook Islands, it is exploited for its gonads by women and children, particularly in the summer months. The animals re-grow their organs, so this harvesting is renewable. This species is also fished for subsistence in Samoa and Tonga. In Asia, it is fished in China, Malaysia, Thailand, Indonesia, the Philippines and Viet Nam. In Southeast Asia, it is known to be part of the "worm" sea cucumbers, lower-value higher-volume species. Also fished in Madagascar.

Regulations: Where it is fished, there are few, or no, regulations pertaining to the harvesting of this species.

Human consumption: Mostly, the reconstituted body wall (bêche-de-mer) is consumed by Asians. The whole animal or its intestine and/or gonads may be consumed as a delicacy or as protein in traditional diets or in times of hardship (i.e. following cyclones).

Main market and value: Singapore and Ho Chi Minh City (Viet Nam) for further exports to Chinese markets. In Viet Nam, it is sold for USD1.3 kg^{-1} dried. It has been traded recently at about USD5 kg^{-1} dried in the Philippines.

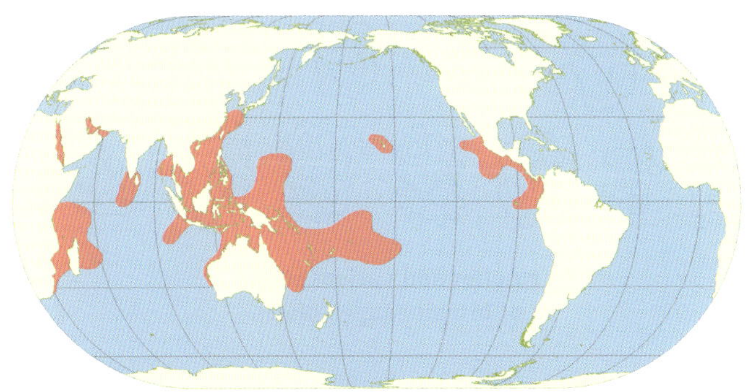

GEOGRAPHICAL DISTRIBUTION:

This species has one of the broadest distributions of all holothurians, and it can be found in most tropical localities in the western central Pacific, Asia and most Indian Ocean regions.

COMMON NAMES: Pepino de mar (Latin American countries), Donkey dung sea cucumber (Bahamas, Florida Keys).

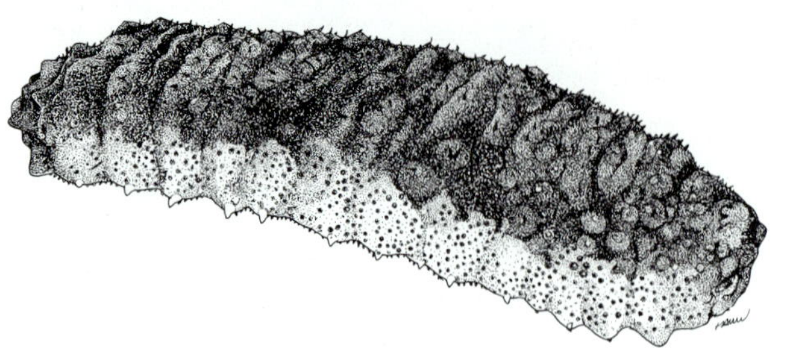

DIAGNOSTIC FEATURES: Dark brown, grey or black dorsally, usually becoming lighter on the lower margins. The dorsal surface is smooth and possesses bumpy, wart-like protrusions. Ventral surface varies greatly from bright red, pink, orange, white, yellowish, grey, dark purple or black. Smallest juveniles are yellowish-white with flecks of purplish-brown on dorsal papillae and few on ventral surface. *Holothuria mexicana* is a large species with a rigid body, with large dorsal and lateral folds. Both anterior and posterior ends are rounded. The ventral mouth contains 20–22 peltate tentacles. The dorsal podia tend to hold detritus, seagrass blades and algae.

Ossicles: Tentacles with rods of various sizes, 55–190 µm long as well as rosettes. Dorsal body wall with tables and rosettes. Tables with irregular discs that have spiny extensions, 50–95 µm across, and are perforated by 4 large central holes and few peripheral ones; spire ending in a spiny crown. Rosettes can be open or closed, forming biscuit-shaped ossicles, 25–50 µm long. Ventral body wall with similar rosettes and fewer tables with discs 40–75 µm across.

Processed appearance: Dark brown to blackish in colour. The body is tapered gradually at both end and possesses a bumpy texture.

Remarks: Studies have shown that this species may accumulate high levels of trace metals including copper, nickel, lead and zinc. The concentrations of these metals were found in the eviscerated digestive tract.

Size: Maximum length about 50 cm; average fresh length 33 cm. Average fresh weigh about 260 g (Panama).

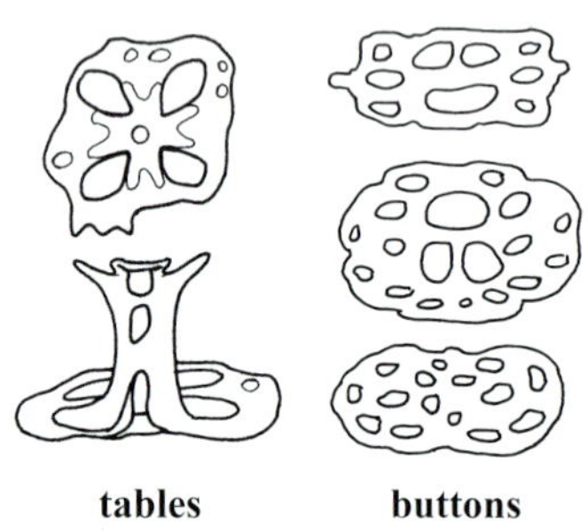

tables   **buttons**

(after Hasbún and Lawrence, 2002)

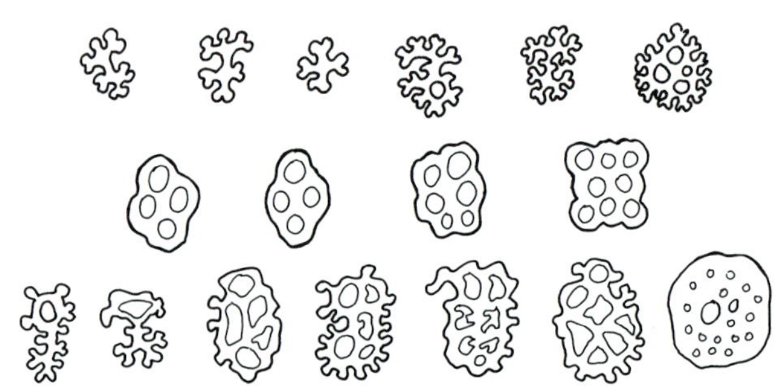

**rosettes and various stages
of biscuit-shaped plates**

(after Deichman, 1957)

HABITAT AND BIOLOGY: In Colombia, this species prefers coral reefs, seagrass beds, sandy or rubble bottoms and mangrove habitats. In the wider Caribbean, it inhabits shallow waters with sandy or coral patches or seagrass beds.

In Panama, it reproduces between February and July, and late summer in southern Florida (USA). However, there are individuals with mature gametes all year long. This species has a size-at-maturity of 18 cm. On Curaçao reefs, it mostly spawns within the first five days following the full moon between August and October.

LIVE (photo by: SIMAC-INVEMAR)

PROCESSED (photo by: F.A. Solís-Marín)

EXPLOITATION:

Fisheries: Harvested by hand collection in artisanal fisheries in the Gulf of Mexico and Caribbean, often illegally. This species is part of multispecies fisheries that often include *H. floridana*, *H. thomasi*, *Astichopus multifidus* and *Isostichopus badionotus*.

Regulations: Where it is fished, there are few, or no, regulations pertaining to harvesting, apart from no-take marine reserves.

Human consumption: Mostly, the reconstituted body wall (bêche-de-mer) is consumed by Asians.

Main market and value: Asian markets. Prices in Hong Kong China SAR retail markets ranged from USD64 to 106 kg^{-1}.

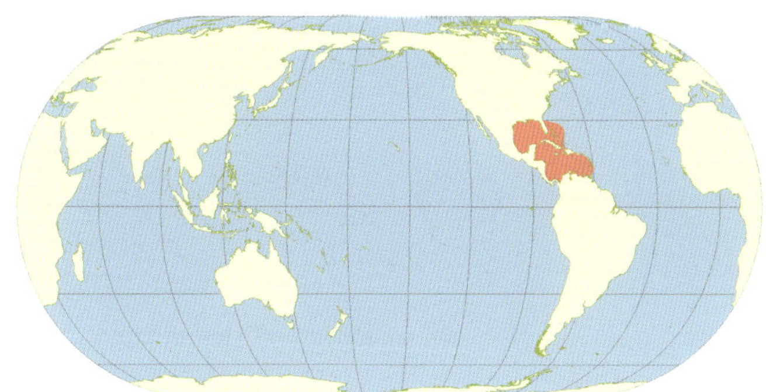

GEOGRAPHICAL DISTRIBUTION: Distributed widely along the Florida Keys, Bahama Islands, Cuba, Puerto Rico, Jamaica, Barbados, Tobago, Aruba, Yucatan Peninsula, Belize, Bonaire, Venezuela (Bolivarian Republic of) and islands off Colombia, at depths from 0.5 to 20 m.

Holothuria nobilis (Selenka, 1867)

COMMON NAMES: Black teatfish (**FAO**), Holothurie noire à mamelles (**FAO**), Benono (Madagascar), Barbara (Mauritius), Bawny black (Egypt), Abu habhab aswed (Eritrea), Pauni mweusi (Kenya and Zanzibar, Tanzania).

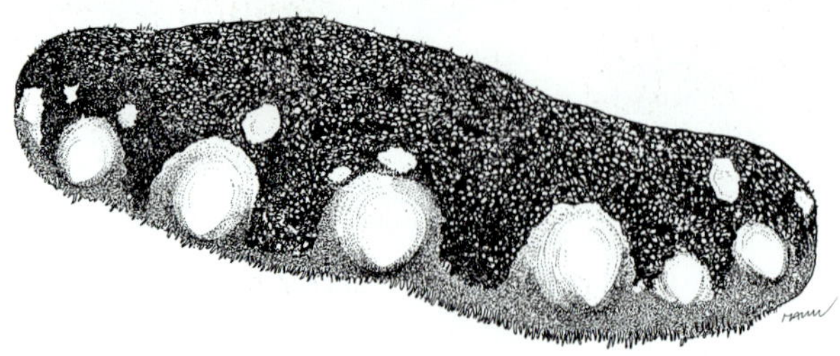

DIAGNOSTIC FEATURES: This species is black dorsally with white blotches and spots on the sides of the animal and around the lateral protrusions ('teats'). Body suboval; stout and very firm, arched dorsally and strongly flattened ventrally. Body wall is thick, and possesses generally 6–10 characteristic large lateral protrusions. The body is often covered by a thin coating of fine sediment. Dorsal podia are sparse and small, while the ventral podia are numerous, short and greyish. The mouth is ventral, with 20 stout tentacles. Anus surrounded by 5 small calcareous teeth. Cuvierian tubules absent. Juveniles probably differ in colour from adults.

Ossicles: Tentacles with rods, 40–410 µm long, spiny at extremities and mostly curved. Dorsal and ventral body wall with the same type of tables. Table discs, 55–70 µm across, circular with an undulating rim, perforated by 4 large central holes and 8–12 peripheral ones; spire low ending in a regular spiny crown or in an irregular one with less spines. Buttons of the dorsal body wall are elongated or ellipsoid, and on average 100 µm long. Buttons of the ventral body wall can be smooth, knobbed, or fenestrated, 80–100 µm long. Ventral and dorsal podia present large perforated plates with ragged sides.

Processed appearance: Processed *Holothuria nobilis* has a flattened, stout shape with obvious teats along both sides of the body. The body surface is powdery greyish-brown, smooth to slightly wrinkled. The ventral body wall is usually dirty grey. One single cut dorsally but not completely to the mouth or anus. Common size 18–24 cm.

Remarks: This species is found on tropical reefs of the western Indian Ocean. The black teatfish of the Pacific is a separate species named *H. whitmaei* Bell, 1887.

Size: Maximum length about 60 cm; average length about 35 cm. Average fresh weight: 230 g (Mauritius), 800–3 000 g (Réunion), 1 500 g (Egypt); average fresh length: 14 cm (Mauritius), 35 cm (Réunion), 55 cm (Egypt).

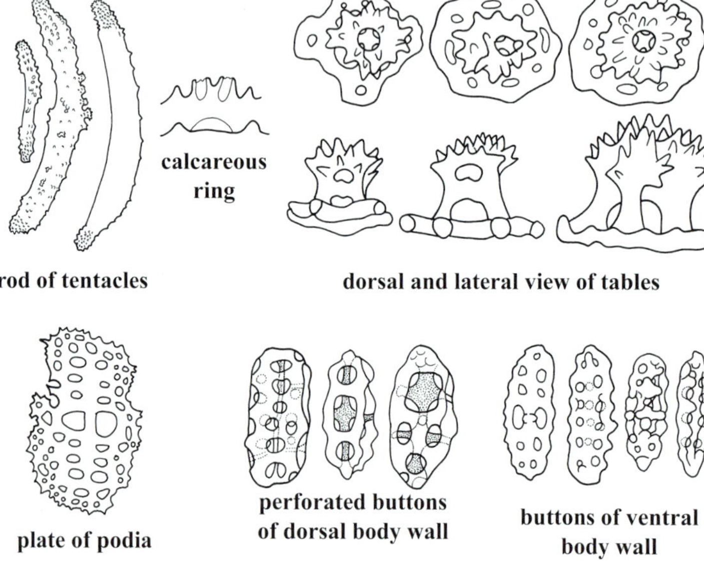

rod of tentacles

calcareous ring

dorsal and lateral view of tables

plate of podia

perforated buttons of dorsal body wall

buttons of ventral body wall

(after Cherbonnier, 1988)

HABITAT AND BIOLOGY: It lives in shallow coral reef habitats (lagoons) up to 20 m depth. In Africa and the western Indian Ocean region, this species can be found on reef flats and slopes on coral rubble between 0 and 40 m depth. In Madagascar, it occurs in the inner slope and on seagrass beds, with higher abundance in the former. In the Comoros, it normally inhabits between 10 and 40 m depth on coarse sand.
It reproduces annually during the cold season.

EXPLOITATION:

Fisheries: *H. nobilis* is one of the most valuable commercial species and, therefore, is overexploited. Exploitation of *H. nobilis* is at scales ranging from artisanal (e.g. Tanzania) to industrial (e.g. Mauritius). This species was previously harvested by hand collecting from reef flats in Egypt.

LIVE (photo by: R. Aumeeruddy)

PROCESSED (photo by: S.W. Purcell)

It is collected by free diving and SCUBA diving in Madagascar and Mauritius. It has been fished commercially in Eritrea, Madagascar, Egypt, Maldives, Mozambique and Seychelles. In Kenya and Tanzania, it is among the most valuable commercial species; however, in Tanzania it is captured in low numbers due to its scarcity. This species has been depleted in Mozambique, India, Madagascar, Egypt, Red Sea, Maldives and probably in Tanzania and Kenya due to overfishing.

Regulations: It is currently banned in Egypt. There is no management for exploitation of this species in Mauritius.

Human consumption: Mostly, the reconstituted body wall (bêche-de-mer) is consumed by Asians.

Main market and value: Markets are Hong Kong China SAR, Singapore, Taiwan Province of China, China and Malaysia. It is sold at USD20–80 kg^{-1} dry wet, depending on size and condition. Prices in Hong Kong China SAR retail markets ranged from USD106 to 139 kg^{-1} dried.

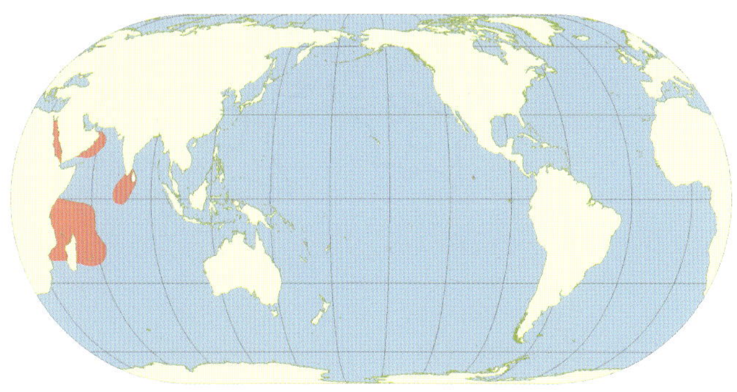

GEOGRAPHICAL DISTRIBUTION: Known from localities in the western Indian Ocean, from East Africa to possibly India and Maldives. It can also be found in the Red and Arabian Seas. This species does not appear to occur as far east as the Java Sea (e.g. western Indonesia) and south China Sea (e.g. Malaysia, Viet Nam, Philippines).

Holothuria notabilis Ludwig, 1875

COMMON NAMES: Dorilisy, Tsimihoke (Madagascar).

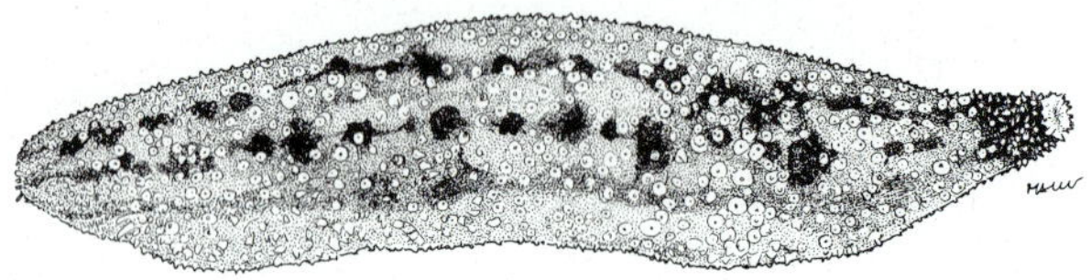

DIAGNOSTIC FEATURES: *Holothuria notabilis* is a medium-sized species. It is cylindrical in shape and tapered near the anus. The mouth is ventral and possesses 20 small yellow tentacles. From the original taxonomic description, the tegument colour is whitish with many dark-brown or black 'dots' on the dorsal surface, forming two rows of 8–10 dots. Podia on the ventral surface are small and scattered over the entire area. The calcareous ring is well developed. It has short white Cuvierian tubules that are rarely expelled.

Ossicles: Numerous small nodulous buttons and few tables with their disc irregularly spined and with spire reduced to 4 short pillars that are fused at their base. Such small, reduced tables are present in the upper tissue of the ventral body wall. In the dorsal body wall tables with larger disc diameter and with fully developed spire. Between both type of tables all intermediates can be found, both in diameter of table disc and in height of spire.

Processed appearance: The specimens look like small wooden sticks. The conversion factor from live to dry is low, at around 3% of original whole body weight.

Size: Average fresh weight: 180 g; average fresh length: 18 cm. Maximum weight is about 500 g and maximum length is 32 cm.

HABITAT AND BIOLOGY: In Madagascar, this species prefers lagoons and seagrass beds on sandy substrata between 0 and 10 m depth. It can be found at densities of about 200 ind. ha^{-1} (with a biomass in fresh weight of about 30 kg ha^{-1}).

The size at first sexual maturity is 9 cm in length or 20 g gutted weight (or 60 g total body weight). The gonads are long branched tubules; at maturity, they are white in males and orange in females. In Madagascar, the annual reproductive cycle annual is well marked, with spawning in November–December.

button

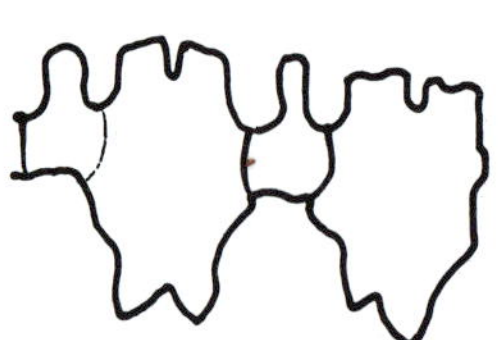

calcareous ring

tables

(after Ludwig, 1875)

EXPLOITATION:

Fisheries: *H. notabilis* entered the trade in Madagascar in 2002 and is currently exploited intensively. It is collected by women and children by wading at low tide and by men snorkeolling at coastal sites from canoes. The catch per unit of effort (CPUE) in Madagascar has been reported at 3 kg day^{-1} fisher^{-1}. This species is fished together with other medium-sized or small species.

Regulations: None except for moratoria within multispecies fisheries in the Western Indian Ocean.

Human consumption: This species is processed into the dried bêche-de-mer product for export and is not consumed locally.

Main market and value: The products from Madagascar are shipped with the other species to the markets of Hong Kong China SAR, China and Singapore.

LIVE (photo by: IH-SM-WIOMSA)

PROCESSED (photo by: IH-SM-WIOMSA)

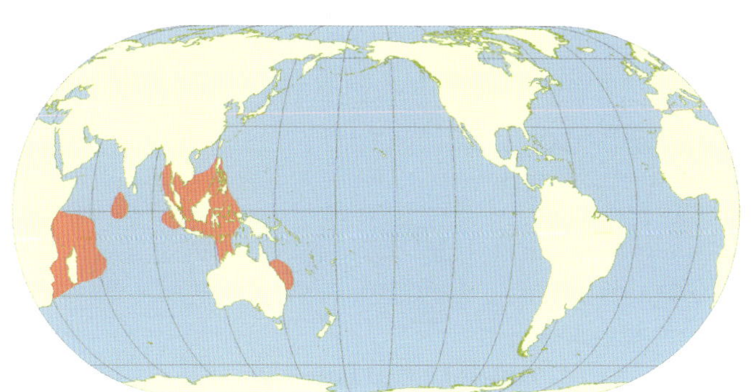

GEOGRAPHICAL DISTRIBUTION:
Great Barrier Reef (Australia) and found at localities in the Indian Ocean including Madagascar, Mozambique, and eastern Indonesia.

COMMON NAMES: Flower teatfish, Pentard (Seychelles), Pauni kaki (Zanzibar, Tanzania).

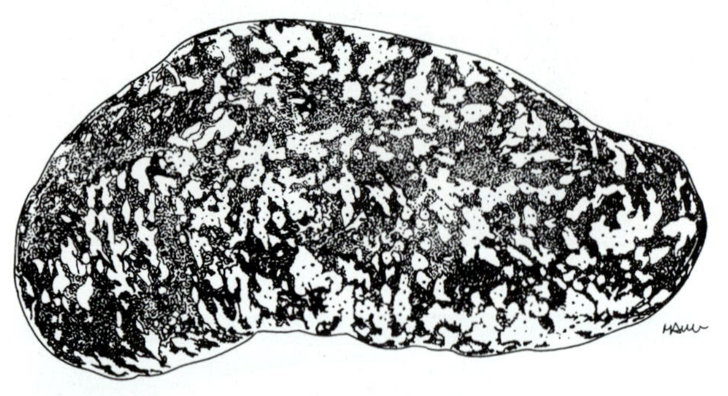

DIAGNOSTIC FEATURES: Dorsal surface is dark brown and mottled with irregular-shaped, cream coloured, blotches. Stout body; arched dorsally and flattened ventrally.

Ossicles: Tentacles with spiny rods, 70–615 µm long. Dorsal body wall with tables, buttons and simple ellipsoid buttons. Tables with round, smooth discs, 60–75 µm across, perforated by a single central hole and one ring of peripheral holes; spire wide ending in a wide crown of spines. Buttons can be smooth or with just some medium-sized knobs with 4–8 pairs of holes, or can be modified into simple ellipsoid buttons, 80–115 µm long. Ventral body wall with tables of roughly the same form and size and with buttons, 110 µm long, that are smoother slightly knobbed and have 4–7 pairs of holes. Ossicles of the podia are, at present, undocumented.

Processed appearance: Whitish body colour when skin removed. The lateral protrusions ('teats') are visible even when dried. The mottling on the dorsal surface remains visible after processing. Common dried size 17 cm.

Remarks: Not yet described taxonomically. Future studies will help decide if it is another species or simply a variety of the Indian Ocean black teatfish. Very little is known apart from its presence in the catches. Catch weights are sometimes combined with those of *Thelenota ananas* in Seychelles.

Size: Average fresh length is 30 cm. Average fresh weight is about 1 675 g.

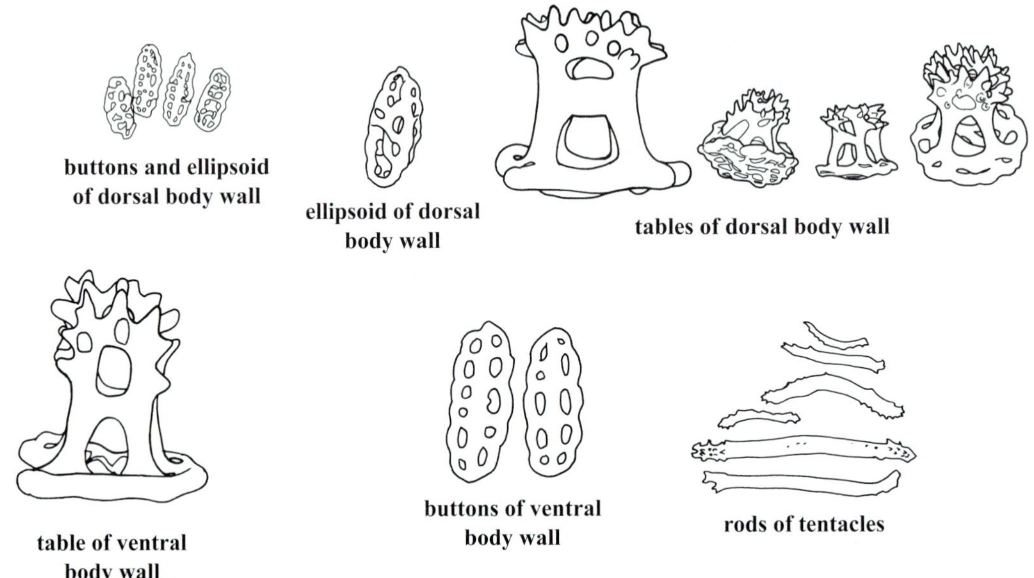

(source: photos D. VandenSpiegel)

HABITAT AND BIOLOGY: In Seychelles, this species prefers lagoons over sandy bottoms between 10 and 50 m deep. Its reproductive biology is unknown.

EXPLOITATION:

Fisheries: Exploitation of this species is at artisanal and semi-industrial scales. *Holothuria* sp. (type 'Pentard') is the main species in trade in Seychelles where it is considered the highest value species. It is harvested by SCUBA diving. This species is part of a multispecies fishery that includes *H. nobilis*, *H. fuscogilva* and *T. ananas*.

Regulations: This fishery is managed in Seychelles by means of a restricted number of fishing permits and no-take reserves.

Human consumption: Mostly, the reconstituted body wall (bêche-de-mer) is consumed by Asians.

Main market and value: Hong Kong China SAR. It is sold at USD17–26 kg^{-1} dried. Retail prices in Hong Kong China SAR were up to USD188 kg^{-1} dried.

LIVE (photo by: R. Aumeeruddy)

PROCESSED (photo by: C. Conand)

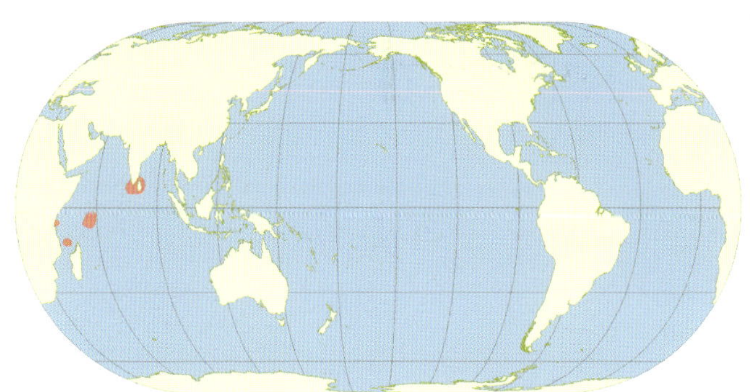

GEOGRAPHICAL DISTRIBUTION:

Known from the Comoros, Nosy Be Island (Madagascar), Seychelles, Zanzibar (Tanzania), Maldives and Sri Lanka, where it is exploited.

COMMON NAMES: Sea cucumber, Bantunan (Indonesia).

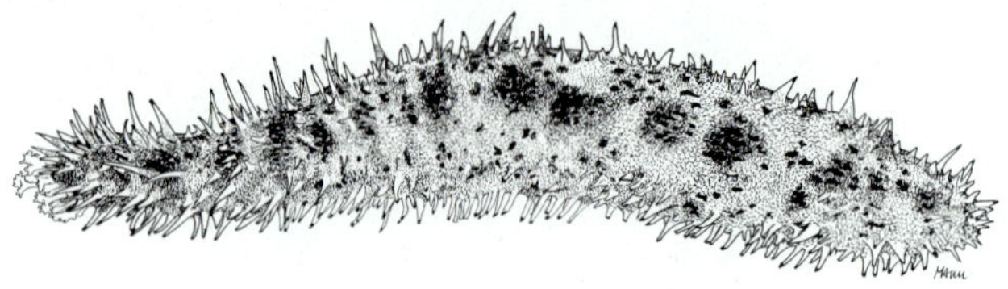

DIAGNOSTIC FEATURES: Body is beige to light yellowish or grey in colour, with 2 rows of large dark spots and numerous tiny dark spots. The body is covered with numerous dark brown or black, short, conical papillae with rounded or slightly conical tips, which are scattered on the dorsal surface. The ventral surface is yellowish to light brown. The body is elongate, cylindrical, and wider at the posterior end of the animal. Ventral podia are short, stout and numerous. The mouth is ventral to terminal, surrounded by a double circle of papillae, and has 18–22 tentacles. The anus is terminal and surrounded by conical papillae. No Cuvierian tubules.

Ossicles: Tentacles with rods up to 180 μm long. Dorsal and ventral body wall with similar tables and buttons. Table discs 50–80 μm across, with smooth or spiny rims, and are perforated by 4 central holes and 4–12 peripheral ones, and the spire ends in a small spiny crown. Buttons, 40–70 μm long, with 3–10 holes, rather irregular. Ventral podia with tables and buttons similar to those of the body wall, and there are perforated plates, up to 150 μm long. Dorsal papillae with tables and buttons similar to those of the body wall and characteristic large, slightly curved rods, 90–170 μm long, that are perforated distally.

Processed appearance: Not available. This species may be traded mixed with other low-value species in the dried form.

Size: Average body length roughly 12 to 25 cm.

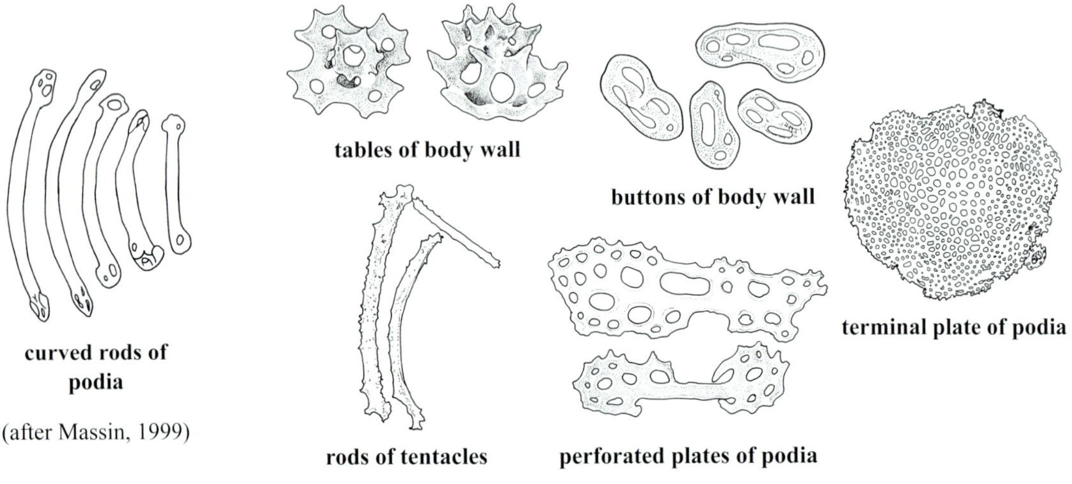

(after Massin, 1999)

(source: Solís-Marín *et al.*, 2009)

HABITAT AND BIOLOGY: In Kenya, it has been observed buried under coral rubble or coral boulders. In the Comoros, it inhabits shallow waters between 0 and 10 m depth on coral rock or buried among coral rubble. In La Réunion, it is found in crevices on reef flats. Its reproductive biology is unknown.

EXPLOITATION:
Fisheries: It is commercially exploited in China and Indonesia.
Regulations: Not available.
Human consumption: Unknown, probably exported dried and eaten by Asians after being reconstituted.
Main market and value: Not available.

LIVE (photo by: www.noaa.gov)

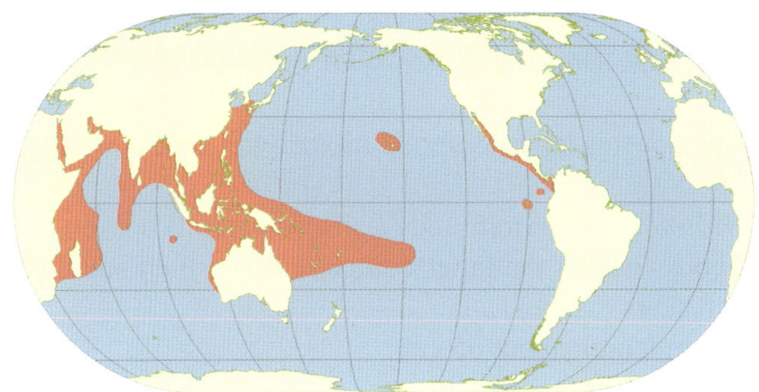

GEOGRAPHICAL DISTRIBUTION: Ranges from the western central Pacific to the Hawaiian Islands, Asia and the Africa and Indian Ocean region. Also found on the Pacific coast of Central America.

COMMON NAMES: Unknown.

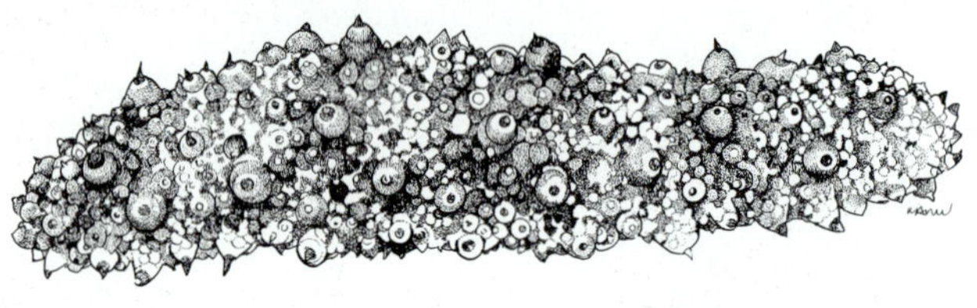

DIAGNOSTIC FEATURES: This species is grey to brown with four to six dark brown blotchy bands across the body. The dorsal surface has relatively sparse but large, yellowish protuberances ending in a brown-coloured papilla with whitish tip. The ventral surface is white to yellowish and is covered with numerous, long, cylindrical podia. The mouth is ventral, with 20 large, yellowish or greyish tentacles that have small brown spots. The anus is terminal, relatively large and surrounded by a wide dark brown ring and by five groups of small white papillae. Cuvierian tubules present, lightly bluish, and readily ejected.

Ossicles: Tentacles with rods, 150–375 µm long, the largest ones are slightly spinose distally. Dorsal and ventral body wall with similar tables and pseudo-buttons. Tables with discs 30–40 µm across, rim smooth and undulating, perforated by 4 central holes and 1–3 smaller peripheral holes; spire, if present, is low and ends in an ill-formed crown. Pseudo-buttons of dorsal body wall are 40–55 µm long, while those of ventral body wall are slightly smaller, 25–35 µm long. Ventral podia with buttons, perforated plates and perforated rods. Dorsal papillae with rods that can turn into perforated plates, up to 190 µm long.

Processed appearance: Not available. This species may be traded mixed with other low-value species in the dried form

Remarks: The external distinction between _Holothuria pervicax_ and _H. fuscocinerea_ is not too easy for the untrained eye, but the ossicles of the body wall of the latter comprise true, albeit small and rather irregular, buttons whereas _H. pervicax_ presents rosettes.

Size: Small to moderate-sized species. Maximum length about 35 cm.

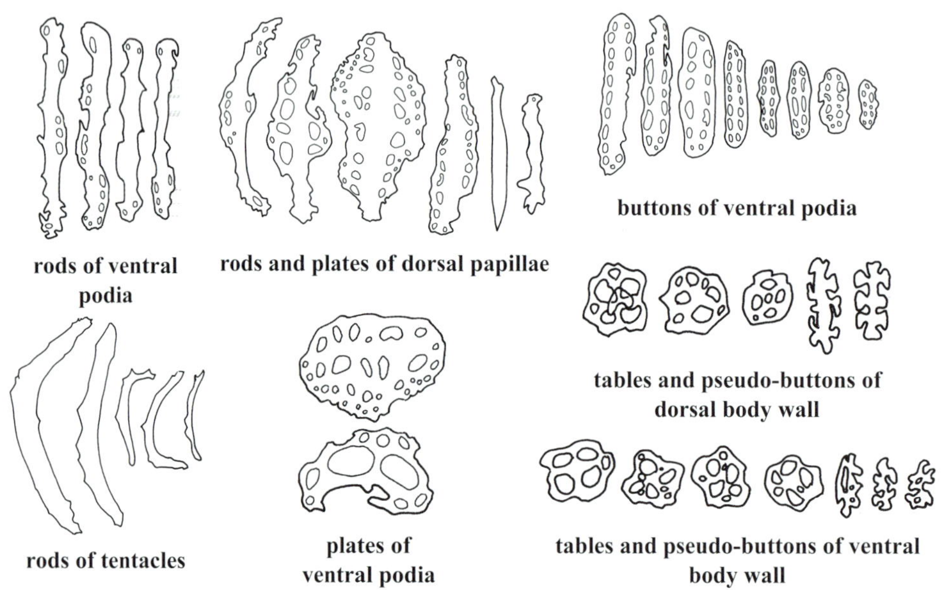

rods of ventral podia

rods and plates of dorsal papillae

buttons of ventral podia

rods of tentacles

plates of ventral podia

tables and pseudo-buttons of dorsal body wall

tables and pseudo-buttons of ventral body wall

(source: Samyn, 2003)

HABITAT AND BIOLOGY: This species can be found on reef flats underneath coral rocks, where it generally remains hidden during the day. It occurs in shallow waters to about 10 m depth.
Its reproductive biology is unknown.

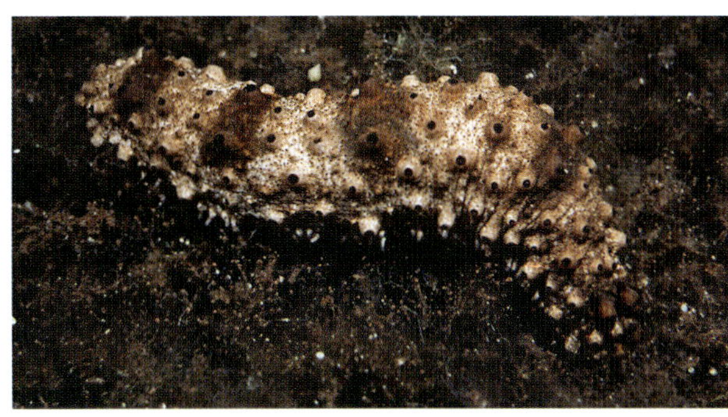
LIVE (photo by: K. Stender)

EXPLOITATION:
Fisheries: This species is known to be fished in China, Madagascar and Indonesia; in the latter, it is part of a multispecies fishery, where *H. pervicax* is used as a filler to top up weights during sales.
Regulations: There appear to be few regulations on the exploitation of this species.
Human consumption: Poorly known, but probably just eaten as reconstituted bêche-de-mer.
Main market and value: Not available. This species is a low-value species. It has been traded recently at about USD3 kg^{-1} dried in the Philippines.

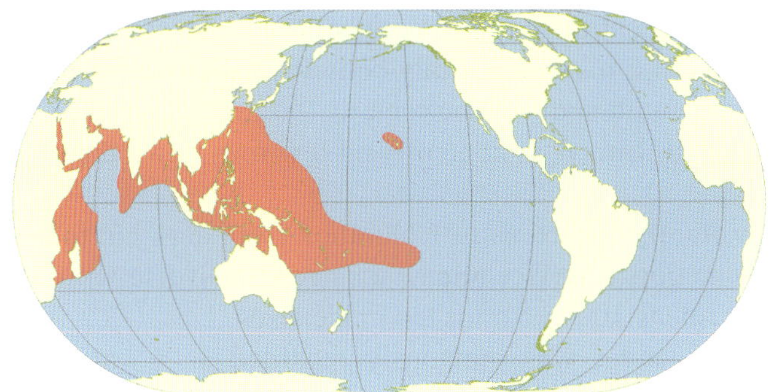

GEOGRAPHICAL DISTRIBUTION:
It can be found widely from localities from the Indian Ocean, Southeast Asia and the Pacific, including Hawaii (USA).

Holothuria scabra Jaeger, 1833

COMMON NAMES: Sandfish (**FAO**), Sand (Egypt), Ñoät traéng (south Viet Nam), Hải sâm trắng, Hải sâm cát (central Viet Nam), Vella attai, Cheena attai (India), Sandfish (Mauritius, Papua New Guinea, Australia), Putian, Cortido, Curtido, Kagisan (Philippines), Hedra beyda (Eritrea), Zanga fotsy (Madagascar), Jongoo mchanga (Tanzania), Myeupe (Zanzibar, Tanzania), Dairo (Fiji), Le gris (New Caledonia).

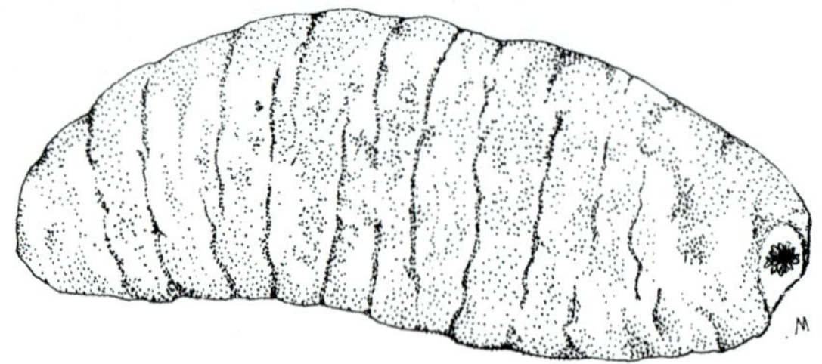

DIAGNOSTIC FEATURES: Colour variable; in the Pacific Ocean and Southeast Asia, it can be black to grey or light brownish green, sometimes with greyish-black transverse lines. In the Indian Ocean, it is usually dark grey with white, beige or yellow transverse stripes. Ventral surface is white or light grey with fine, dark spots. Body oval; arched dorsally and moderately flattened ventrally. Dorsal surface with deep (3 mm) wrinkles and short (1.5 mm) papillae. Body is often covered by fine muddy-sand. Mouth is ventral with 20 small, greyish, tentacles. Anus is terminal with no teeth. No Cuvierian tubules.

Ossicles: Tentacles with spiny rods, 80–440 µm long, slightly curved. Dorsal and ventral body wall with tables and buttons. Ventrally body wall: tables are rare, disc between 60 and 95 µm across, quadrangular and with smooth rim, perforated by 1 central and 8–16 peripheral holes, spire ending in crown of blunt spines; numerous buttons are 40–75 µm long. Dorsally body wall, similar tables, but smaller; buttons are 40–50 µm long. Ventral podia with nodulous buttons, 40–90 µm long, perforated rods, 110–170 µm long, and tables as in body wall. Dorsal papillae present few rods, few tables, but many buttons as those in the body wall.

Processed appearance: Cylindrical with bluntly curved ends. Coloration from dark tan to near black; ventral surface usually amber-brown. The dorsal surface retains the deep transverse wrinkles. No cuts or small cuts across mouth. Dried specimens 10–15 cm for top grade sizes.

Remarks: The golden sandfish, *Holothuria lessoni*, was previously regarded as a variety of *H. scabra*. Recent integrative taxonomic study (Massin *et al.*, 2009) has however clearly separated both species. Whether the Indian Ocean form of *H. scabra* also represents a separate species remains to be investigated.

Size: Maximum length about 40 cm; average length about 24 cm. Maximum weight 2.0 kg; average fresh weight: 300 g (Papua New Guinea, Oman, India), 335 g (Australia), 500 g (Egypt), 580 g (New Caledonia); average fresh length: 19 cm (Australia), 20 cm (New Caledonia, Oman), 22 cm (Papua New Guinea), 25 cm (India), 37 cm (Egypt).

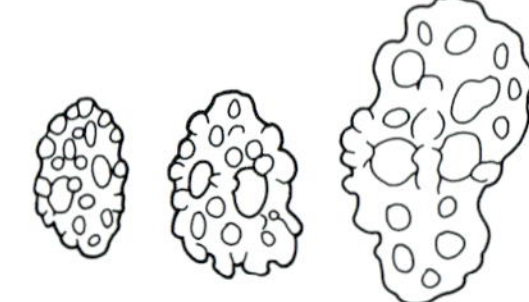

buttons of dorsal body wall

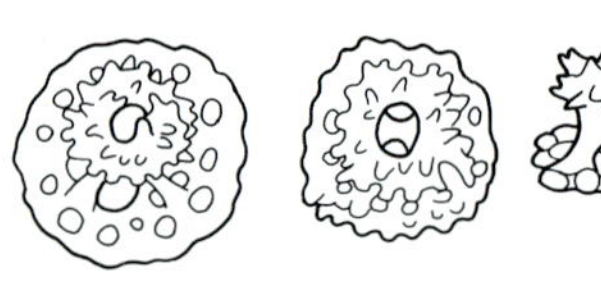

tables of dorsal body wall

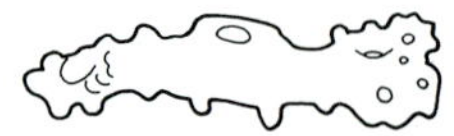

rod of podia

(after Cherbonnier, 1980)

HABITAT AND BIOLOGY: Found in shallow waters, but occasionally to about 20 m. Commonly found on inner flat reefs of fringing and lagoonal reefs, and coastal sandflats and seagrass beds with muddy sandy substrates, near mangroves. Both adults and juveniles bury in sand and sandy-mud at some localities. Attains size-at-maturity at 21 cm in Mauritius, at about 25 cm in India and northern Australia and at 16 cm in New Caledonia.

EXPLOITATION:

Fisheries: Exploited heavily in artisanal (e.g. Madagascar, New Caledonia, Oman, Viet Nam) and industrial fisheries (e.g. Australia, Mauritius). Harvested by free diving and by wading on reef flats in many locations, and by SCUBA or hookah in some localities. In the Western Pacific, it is commercially exploited in practically all localities, east to Fiji. Overfished in Papua New Guinea, Solomon Islands and Vanuatu, leading to moratoria. It is the predominant species fished in the Northern Territory and Western Australia. In Asia, *H. scabra* is harvested in China, Japan, Malaysia, Thailand, Viet Nam, Indonesia and the Philippines. In Africa and the Indian Ocean region, it is fished in most countries within its range. In India, it was one of the most important species for the last 1000 years, until a fishing moratorium in 2001.

Regulations: Before a moratorium in Papua New Guinea, regulations included minimum size limits; 22 cm live and 10 cm dry. Minimum size limits are 20 cm live or 10 cm dried in New Caledonia, 18 cm in Torres Strait, 17 cm in Moreton Bay (Australia), and 16 cm in Great Barrier Reef, Northern Territory and Western Australia. In Fiji, it is banned for export to preserve local consumption. In Oman, SCUBA diving gear is prohibited, there is a closed season and limited access.

Human consumption: Mostly, the reconstituted body wall (bêche-de-mer) is consumed by Asians and is highly regarded. In some parts of the Western Pacific (e.g. Fiji), it is cooked and consumed whole in traditional diets or in times of hardship (i.e. following cyclones). In Viet Nam, it is used commercially for the preparation of traditional medicinal products.

Main market and value: Main market hubs are United Arab Emirates, China and Singapore. It is sold at USD33–47 kg^{-1} dry in Viet Nam and USD0.8 per piece fresh and USD90 kg^{-1} dry in Oman and USD42–88 kg^{-1} dry in the Philippines. It is the most valuable species (by dried) in New Caledonia, exported for USD60–110 kg^{-1} dried and fishers receive USD4–6 kg^{-1} wet (gutted) weight. In Fiji, fishers receive USD3 per piece fresh, even though exports are banned. In Australia, it is sold by fishers for USD6–8 kg^{-1} wet (gutted) weight. Prices in Hong Kong China SAR retail markets ranged from USD115 to 1 668 kg^{-1}. Prices in Guangzhou wholesale markets ranged from USD108 to 200 kg^{-1} dried.

LIVE (Pacific Ocean variety) (photo by: S.W. Purcell)

PROCESSED (Pacific Ocean variety)
(photo by: S.W. Purcell)

GEOGRAPHICAL DISTRIBUTION:

Widespread in the tropical Indo-Pacific, excluding Hawaii, between latitudes 30°N and 30°S and not found further east than Fiji.

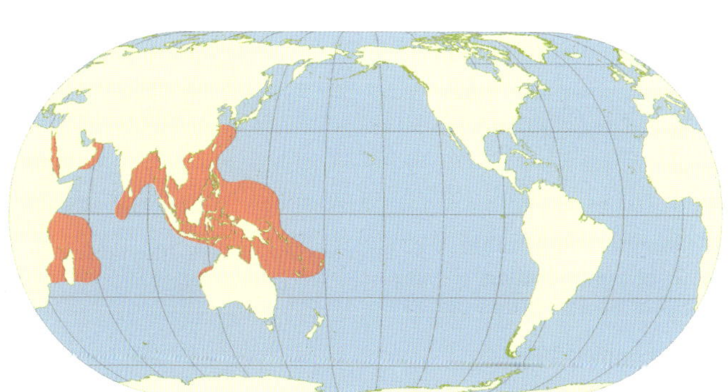

Holothuria spinifera Théel, 1886

COMMON NAMES: Brownfish, Raja attai, Cheena attai (India), Galatta or Weli-atta (Sri Lanka), Nanasi (Zanzibar, Tanzania).

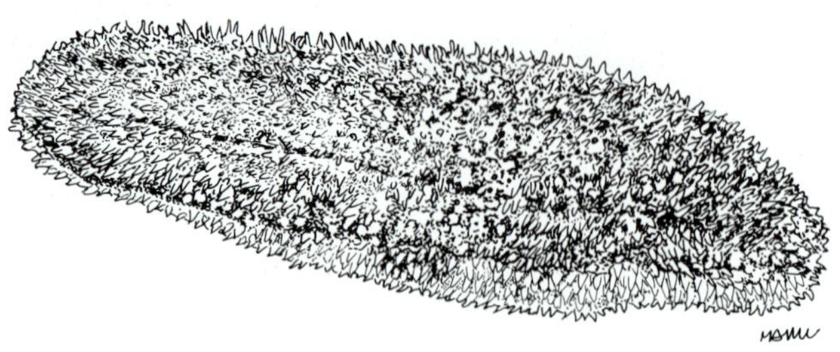

DIAGNOSTIC FEATURES: Dorsal surface brown, becoming lighter on the ventral surface. This species has numerous, small, pointy papillae over the entire body. **Ossicles:** Tentacles with long rods with spiny extremities, up to 500 µm long. Dorsal and ventral body wall with the same types of tables and buttons. Buttons are very nodulous, generally with 3 pairs of holes, but other types also present. Tables present a well developed disc, perforated by 4 central holes and a circle of peripheral ones; spire quite stout and low, ending in an open crown of spines. Ventral podia with perforated plates that may be expanded centrally. Characteristic for this species are the large tack-like tables of the dorsal/anal papillae, which can be up to 200 µm high.

Processed appearance: Cylindrical in shape. Dorsal surface is rough, light brown; ventral surface is smooth, light brown. Small cut at the posterior end. Common dried size 8–10 cm.

Size: Maximum length about 30 cm. Average fresh weight: 300 g; average fresh length: 30 cm.

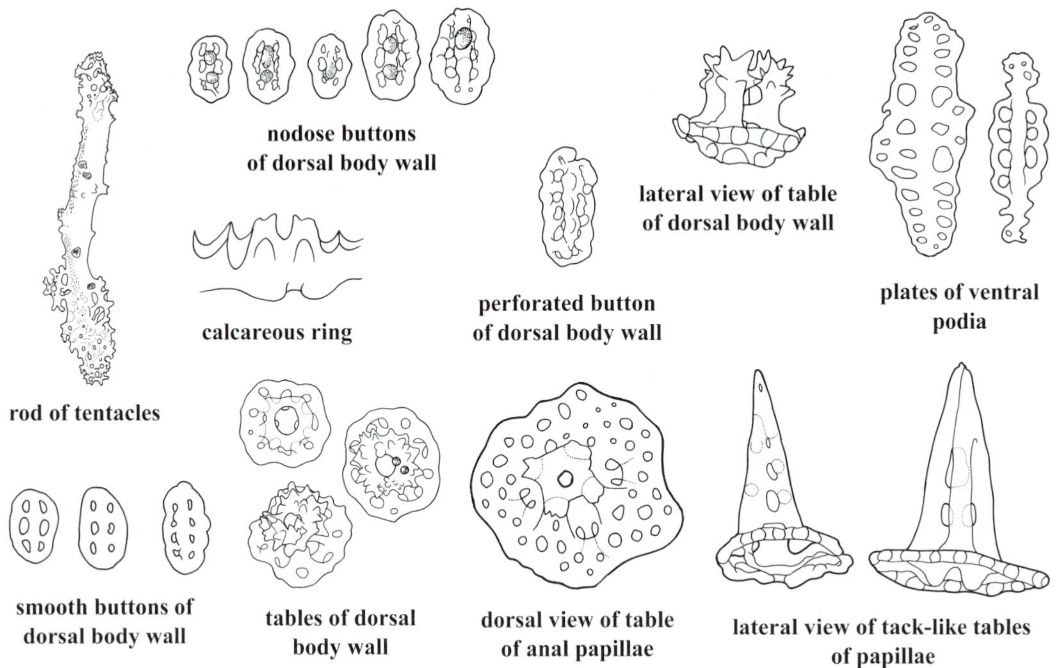

nodose buttons
of dorsal body wall

lateral view of table
of dorsal body wall

plates of ventral
podia

calcareous ring

perforated button
of dorsal body wall

rod of tentacles

smooth buttons of
dorsal body wall

tables of dorsal
body wall

dorsal view of table
of anal papillae

lateral view of tack-like tables
of papillae

(after Cherbonnier, 1955)

HABITAT AND BIOLOGY: *Holothuria spinifera* can completely bury itself in sand in shallow waters from 2 to 10 m. Believed to never be encountered in the intertidal region. It reproduces bi-annually as it has a major peak in September and October and a minor peak in February and March. In Tuticorin (India), it has a prolonged spawning event from November to March.

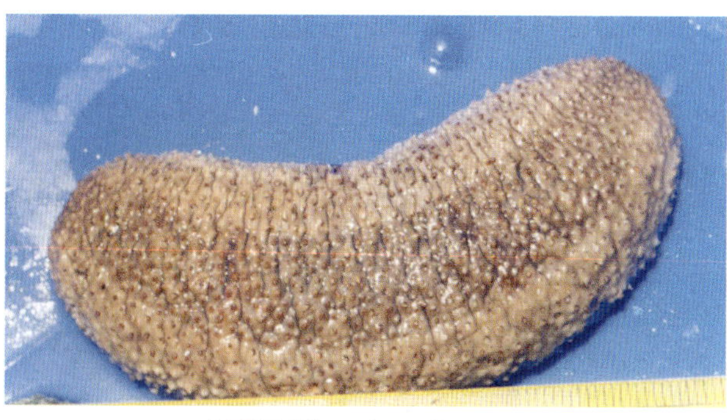

LIVE (photo by: P.S. Asha)

EXPLOITATION:

Fisheries: Previously fished in India by diving and trawling. It is fished in Sri Lanka and some other islands of the Indian Ocean. In Tanzania, it is considered one of the most valuable species. In India, it has been an economically important species for perhaps 1 000 years.

Regulations: In India, the fishery has been banned since 2001.

PROCESSED (photo by: D.B. James)

Human consumption: Mostly, the reconstituted body wall (bêche-de-mer) is consumed by Asians.

Main market and value: It is a moderate-value species. Prices in Hong Kong China SAR retail markets ranged from USD160 to 188 kg^{-1} dried.

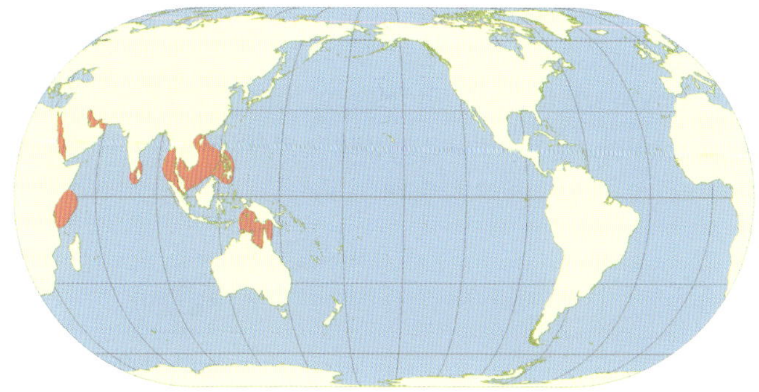

GEOGRAPHICAL DISTRIBUTION:
Red Sea, Persian Gulf, Sri Lanka, northern Australia, the Philippines. In India, it is known only from Gulf of Mannar and Palk Bay.

Holothuria whitmaei Bell, 1887

COMMON NAMES: Black teatfish, Bakungan, Kagisan, Sus-uan (Philippines), Le tété noir (New Caledonia, Wallis and Futuna Islands), Teromamma (Kiribati), Loaloa (Fiji), Huhuvalu uliuli (Tonga), Ñoät ñen ña, Đồn đột vuù (Viet Nam), Susu (Malaysia).

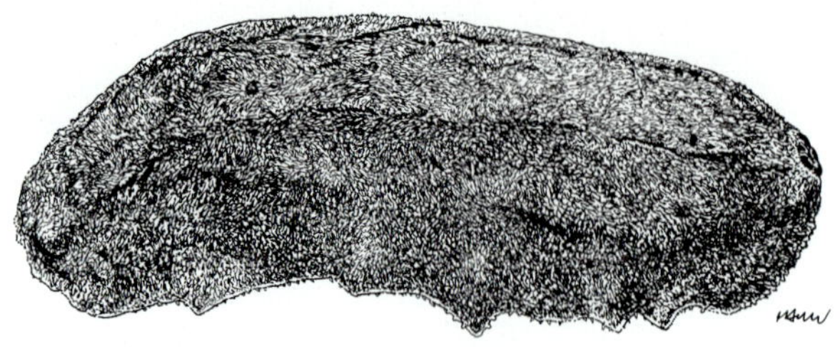

DIAGNOSTIC FEATURES: This species is uniformly black dorsally, and dark grey ventrally. Juveniles may have beige or white markings on the dorsal surface but ventrally are usually dark grey. It possesses 5–10 large stout, pointed, protrusions ('teats') at the lateral margins of the ventral surface, which may retract totally when handled or preserved. Body generally covered by a thin coating of fine sand. Body suboval; stout and very firm; arched dorsally and strongly flattened ventrally and with rounded ends. Body wall is thick. Dorsal podia are sparse and small, while the ventral podia are numerous, short and brown to grey. The mouth is ventral, with 20 stout tentacles. Anus surrounded by 5 small calcareous teeth. Cuvierian tubules are few, short, and not expelled.

Ossicles: Tentacles with rods of various sizes, 100–335 µm long, spiny at extremities, but not branching. Dorsal body wall with tables and ellipsoid buttons. Table discs are 70–85 µm across, perforated by one central and a ring of peripheral holes; the stout, but low, spire ends in a wide spiny crown. Ellipsoid buttons are 50–70 µm long and quite irregular, and perforated by 3–5 holes. Ventral body wall with similar tables as those of the dorsal body wall, and with ellipsoid buttons about 55–85 µm long that are more elongated and with holes more occluded, as well as long buttons that are nearly smooth.

Processed appearance: Processed *Holothuria whitmaei* have a flattened, stout shape with rounded teats along both sides of the body. The body surface is powdery dark grey and relatively smooth. The ventral surface is brownish-grey with fine bumps. One straight cut along the dorsal body wall but not completely to the mouth or anus. Common size 15–20 cm.

Remarks: *H. whitmaei* was considered a synonym of *H. nobilis*. Recent integrative taxonomic research has however shown that both species are valid. *H. whitmaei* is found in the West Pacific and *H. nobilis* in the Indian Ocean.

Size: Maximum length about 54 cm; average length is 34 cm. In New Caledonia, average live weight was recorded at 1 800 g and average live length about 23 cm.

HABITAT AND BIOLOGY: In the western central Pacific this species can be found in reef flats, reef slopes and sandy (but not muddy) seagrass beds between 0 and 20 m. On the Great Barrier Reef (Australia), Western Australia and New Caledonia, it has an annual reproductive event with average gonadal index reaching its maximum between April

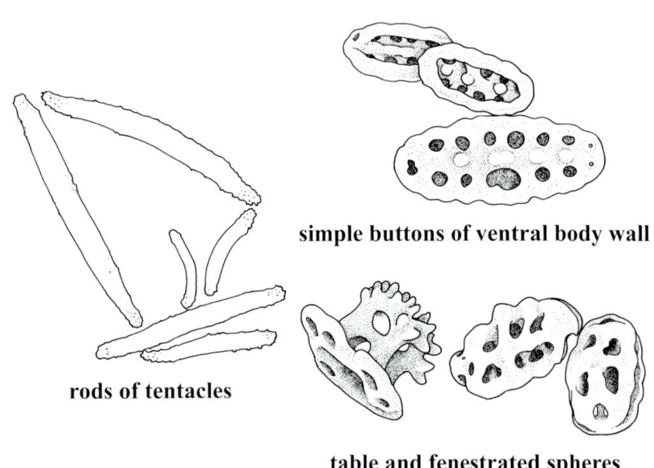

(source: Uthicke *et al.*, 2004)

and June, showing that it is one of the few tropical species that reproduce during the winter.

EXPLOITATION:

Fisheries: In the central western Pacific, this species is commercially exploited in practically all localities within its range where it can be legally fished. It was overfished in Papua New Guinea, Solomon Islands and Vanuatu, leading to nationwide moratoria. There are subsistence fisheries for this species in Palau and French Polynesia. In New Caledonia, it has become the most heavily fished species of sea cucumber. On the Great Barrier Reef, overexploitation reduced stocks on fished reefs by about 80%, leading to a long-term moratorium, and stocks have not yet shown signs of recovery. In Asia, this species is heavily fished in Indonesia, China and the Philippines.

LIVE (photo by: S.W. Purcell)

PROCESSED (photo by: S.W. Purcell)

Regulations: Before the national moratorium in Papua New Guinea, there was a minimum size limit of 22 cm wet or 10 cm dry. Size limits elsewhere are: 30 cm live and 16 cm dry in New Caledonia; 25 cm live in Torres Strait; 30 cm live on the Great Barrier Reef; 26 cm live in the Northern Territory and Western Australia. Due to large declines in abundances, there is a zero quota in place (effectively a single-species moratorium) in the Great Barrier Reef fishery for this species.

Human consumption: Mostly, the reconstituted body wall (bêche-de-mer) is consumed by Asians and is highly regarded. In some Pacific islands, its intestine and/or gonads may be consumed in traditional diets or in times of hardship (i.e. following cyclones).

Main market and value: The main market is China. It is a high-value species. It has been traded recently at USD23–104 kg^{-1} dried in the Philippines. In New Caledonia, fishers receive USD4–6 kg^{-1} wet and it is exported for about USD40–50 kg^{-1} dried. In Fiji, fishers receive USD11–18 per piece fresh. Prices in Hong Kong China SAR retail markets ranged from USD137 to 231 kg^{-1}. Prices in Guangzhou wholesale markets ranged from USD25 to 116 kg^{-1} dried.

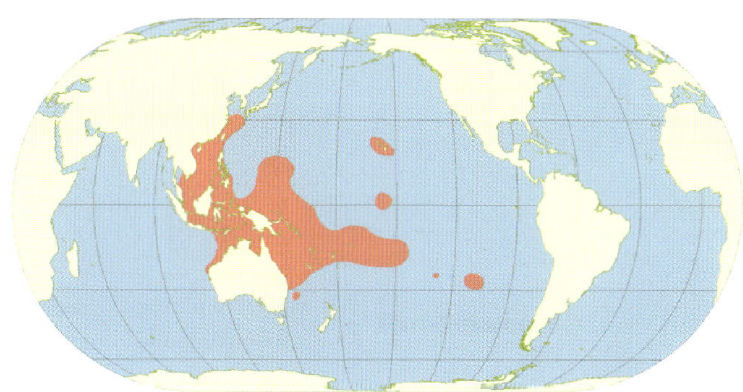

GEOGRAPHICAL DISTRIBUTION:

From Western Australia east to Hawaii and French Polynesia and southern China south to Lord Howe Island, 31°S (Australia). Records of *H. nobilis* from Pitcairn Islands and Easter Island are most probably *H. whitmaei*.

Apostichopus japonicus (Selenka, 1867)

COMMON NAMES: Japanese sea cucumber (**FAO**), Bêche-de-mer japonaise (**FAO**), Cohombro de mar japonés (**FAO**), Cishen (China), Manamako (Japan, also "Aka namako" for red, "Namako" for green and "Kuro namako" for black individuals).

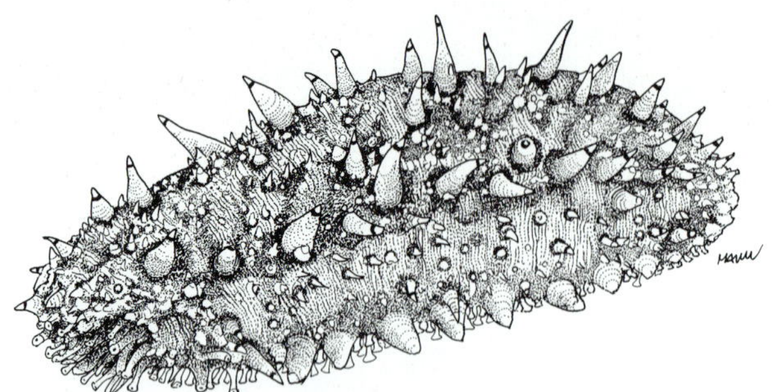

DIAGNOSTIC FEATURES: Dorsal surface variable in colour from brown to grey or olive green; ventral surface brown to grey. Small brownish to greyish dots may be present dorsally, and are more numerous ventrally. Body is squarish in cross-section and tapered somewhat at anterior and posterior ends. Large conical papillae are present in two loose rows on the dorsal surface of the animals and two rows at the lateral margins of the ventral surface. Ventral podia are alined in three irregular longitudinal rows. Mouth is ventral with 20 tentacles. Anus is terminal with no teeth.

Ossicles: Tentacles with curved, spiny rods. In adults, tables are rudimentary (reduced to the spiny disc) or rare in the body wall. Small individuals have their tables better developed. C-shaped rods and rosettes are never present. The cloacal wall has numerous very complex plates. Ventral podia with similar tables as those in the body wall and with simple supporting rods. Dorsal podia with tables with more elaborate spire and perforated supporting rods.

Processed appearance: Cooked and dried animals are dark grey to dark brown and possess characteristic, lighter grey, pointed protrusions in rows along the body.

Size: Average fresh weight: 200 g; average fresh length: 20 cm.

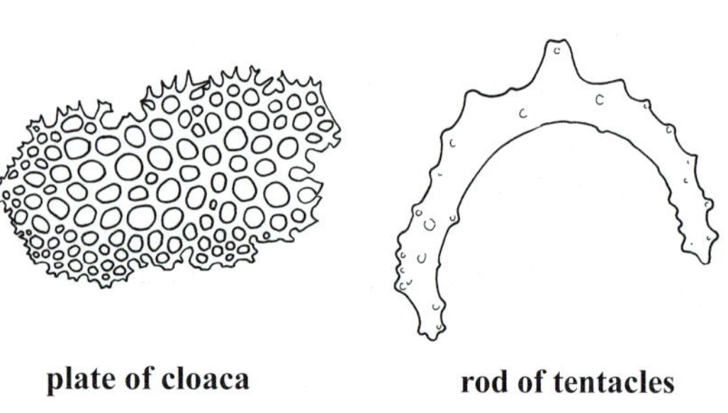

plate of cloaca   rod of tentacles

(source: photo J. Chen)

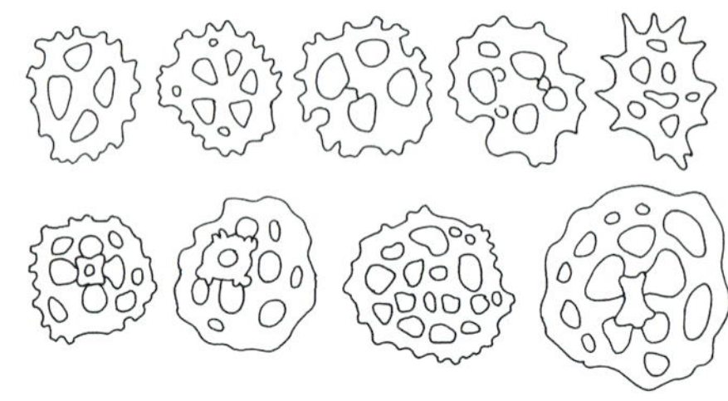

reduced tables of body wall

HABITAT AND BIOLOGY: *Apostichopus japonicus* occurs from the shallows of the intertidal zone to about 20 or 30 m depth. The factors affecting the presence of this species are water temperature, salinity, substratum, attachment sites and habitats for juveniles.

It reaches size-at-maturity at 110 g. This species has an annual reproductive cycle that coincides with the dry season from May to July.

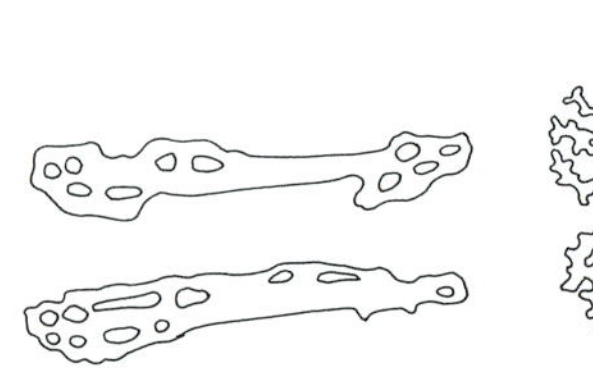

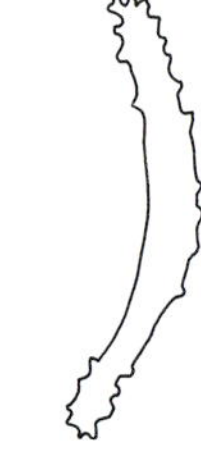

supporting rods   complex plate of cloacal wall   rod of tentacles

(after Liao, 1980)

EXPLOITATION:

Fisheries: This species is fished on an industrial scale, by SCUBA or hookah divers, or by drag nets trawled behind larger boats. It is the most important commercial species in Northeast Asia and has the longest history of exploitation in the Far East (Russian Federation, China, Japan, Republic of Korea and Democratic People's Republic of Korea). China produces about 4 000 tonnes (dried) of *A. japonicus* from aquaculture per year to supply local demand. Japan has the highest fishery captures of this species.

Regulations: In China, exploitation of this species is regulated by fishing permits. Japan has set aside certain localities as breeding reserves where sea cucumber fishing is strictly prohibited, fishing is prohibited during the spawning season from 1 May to 15 June, with a minimum legal weight limit of 130 g.

LIVE (photo by: A. Semenov)

PROCESSED (photo by: S.W. Purcell)

Human consumption: Consumed either as bêche-de-mer, its intestines (konowata) and dried gonads (kuchiko) are eaten as delicacies, or it is eaten raw with sauce. It is commonly used in traditional medicine.

Main market and value: The majority of harvested animals are destined for domestic consumption. It is sold at USD2–3 per unit fresh, USD120–130 kg^{-1} in brine and up to USD400–500 kg^{-1} dried. Prices in Hong Kong China SAR retail markets ranged from USD970 to 2 950 kg^{-1} dried.

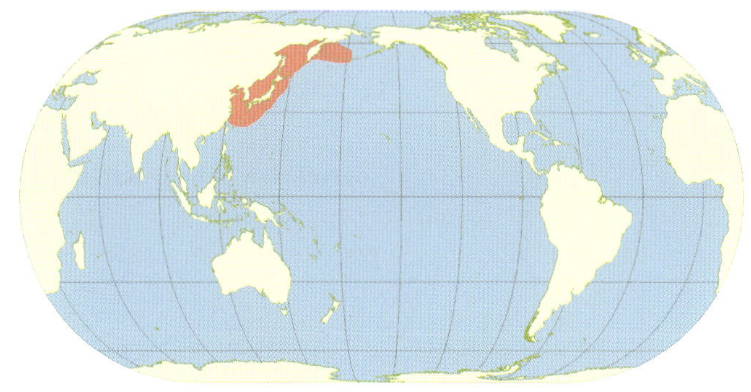

GEOGRAPHICAL DISTRIBUTION:

Distributed mainly in the western Pacific Ocean, the Yellow Sea, the Sea of Japan, the Sea of Okhotsk. The northern limits of its geographic distribution are the coasts of Sakhalin Island, Russian Federation and Alaska (USA). The southern limit is Tanega-shima in Japan. In China, it is commonly distributed on the coast of Liaoning, Hebei and Shandong Province, Yantai and Qingdao of Shandong Province. Its southern limit in China is Dalian Island in Lian Yungang, Jiangsu Province.

COMMON NAMES: Warty sea cucumber (**FAO**), Pepino de mar (Mexico).

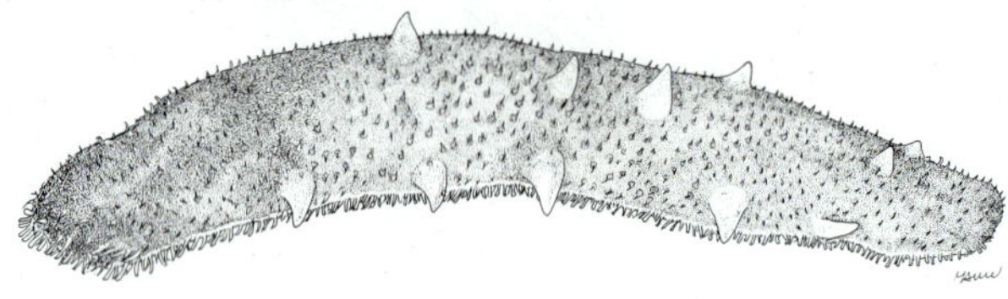

DIAGNOSTIC FEATURES: Coloration is variable from orange to reddish to brownish-grey. In comparison with *Parastichopus californicus*, it is lighter in colour ventrally. Dorsal surface with numerous small, black-tipped papillae that are interspersed with larger, orangish, conical papillae. Body cylindrical. Podia are numerous on the ventral surface. Mouth ventral.

Ossicles: Tentacles with spiny rods, up to 600 µm long. Dorsal and ventral body wall with tables and buttons. Tables with discs only about 45 µm across and rarely with more than 4 perforations; spire ending in a narrow spiny crown. Buttons about 90 µm long and with 3–4 pairs of holes, quite asymmetrical.

Processed appearance: Greyish-brown in colour with a pimply texture to the dried body surface. The large conical papillae are still apparent.

Remarks: See remarks for *P. californicus*.

Size: Maximum length to perhaps 60 cm, but more probably about 40 cm. Average fresh length probably about 30 cm.

HABITAT AND BIOLOGY: This species prefers subtidal habitats on both rocky and soft habitats and is found in low energy environments from the intertidal zone to 30 m depth. It is most abundant in areas with high organic content; small individuals in Santa Catalina (California) reported to feed on fine material from rock rubble under kelp. Juveniles between 2 and 6 cm can be found under rocks, whereas larger juveniles between 8 and 12 cm can be found both on and under rocks. Adults tend to avoid rocks and are found on sediments and feed during both day and night.

The spawning peak is from February to May in Baja California. Weight at first maturity is 140 g at Isla Natividad and 120 g at Bahía Tortuga, Mexico. This species is known to undergo seasonal evisceration, which affects 60% of individuals in October and November.

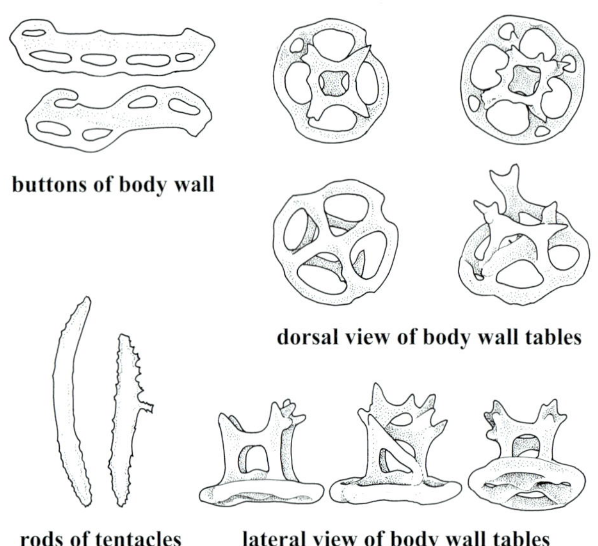

buttons of body wall

dorsal view of body wall tables

rods of tentacles

lateral view of body wall tables

(source: Solís-Marín *et al.*, 2009)

EXPLOITATION:

Fisheries: It is under commercial exploitation on the west coast of Mexico and southern California. Similar to fisheries for *P. californicus*, this species is also exploited by hand collection, by SCUBA diving and by trawling. In California, sea urchin fishers with permits are also allowed to collect sea cucumbers, so it is collected opportunistically in that fishery. Fishing in California began in 1978, whereas it appears that this species was exploited in Mexico only from the late 1980s. The animals are caught in trawls in southern California and by SCUBA divers in northern California. Around islands in southern California, catch rates of several hundred kg fisher^{-1} h^{-1} were reported a decade ago.

Regulations: Within the fishery for this species, there are no-take marine reserves to protect breeding populations. There are permits for each gear type and limited entry restrictions. A TAC regulation is imposed in some fisheries. Trawling is prohibited in some conservation areas.

Human consumption: Mostly, the reconstituted body wall (bêche-de-mer) is consumed by Asians. The muscle strips of *Apostichopus parvimensis* are also consumed.

Main market and value: It is sold by fishers for up to about USD1 kg^{-1} wet, while the processed (dried) animals fetch about USD9 kg^{-1} dried.

LIVE (photo by: J.M. Watanabe)

PROCESSED (photo by: J. Akamine)

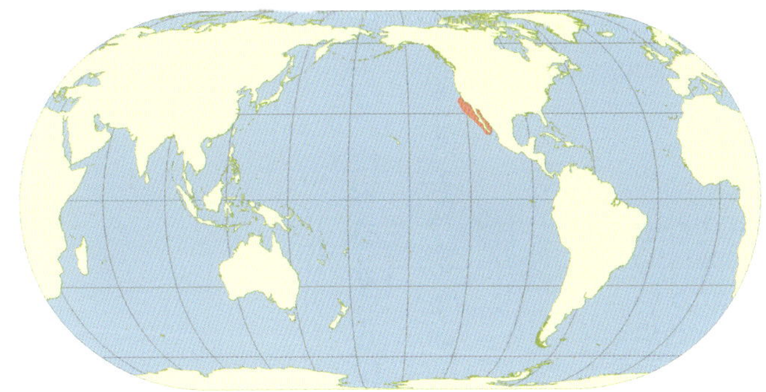

GEOGRAPHICAL DISTRIBUTION:

The distribution of this species is much more restricted, and southern, compared with that of *P. californicus*. It can be found from the Gulf of California to as far north as Point Conception, California.

COMMON NAMES: Furry sea cucumber (Caribbean, Bahamas, Florida, USA), Pepino de mar (Panama).

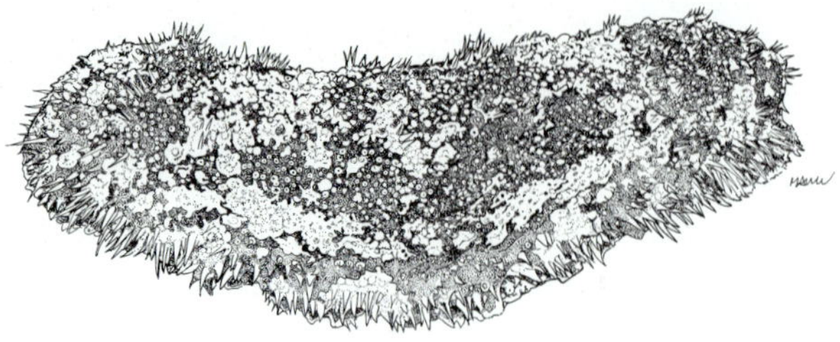

DIAGNOSTIC FEATURES: Dorsal surface brown to yellowish-grey with numerous white spots and white blotches. The ventral surface is usually white, with numerous stout, densely packed podia. Body wall is relatively thick but soft to the touch. The dorsal and lateral surfaces are covered with hundreds of small, conical papillae of 3–5 mm in length; these papillae give the animal a furry appearance. Body highly arched dorsally and flattened ventrally. Mouth is ventral and surrounded by a narrow, yet distinct, tentacular collar. The mouth has 20 tentacles, which have large knob-like discs.

Ossicles: Tentacles of larger specimens with straight or slightly curved rods, up to 175 µm long, which are spinous at the ends and usually also along the shaft. Tentacles also with C-, O- and S-shaped ossicles, about 50 µm long, that are often spiny. The body wall of smaller individuals (less than 20 cm long) has been reported to have large, aberrant tables reminiscent somewhat to those of synallactids. In larger individuals (>20 cm) those tables are missing but numerous C-, O- and S-shaped elements, 40–80 µm long, can be found in the body wall.

Processed appearance: Not available.

Size: Maximum length to at least 50 cm; maximum weight up to 2.5 kg.

HABITAT AND BIOLOGY: It prefers soft bottoms with muddy or sandy patches, in and around seagrass beds mainly of *Thalassia* and *Syringodium*. This species prefers deeper or calmer reef environments in comparison with *Isostichopus badionotus* and *Holothuria mexicana*. It inhabits between 1 and 37 m depth in areas of calcareous algae.

Its reproductive biology is unknown.

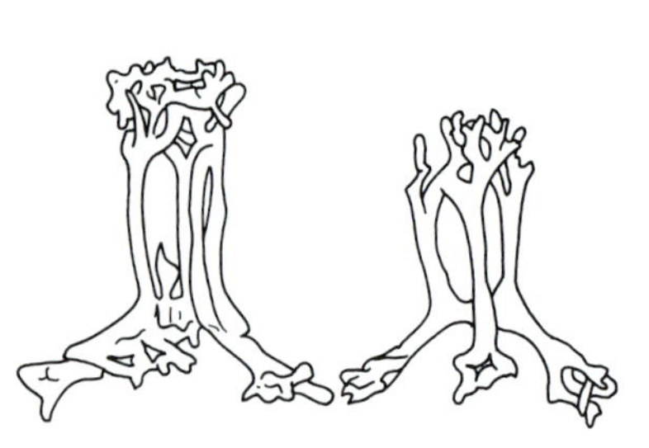

aberrant tables of dorsal body wall

aberrant tables of ventral body wall

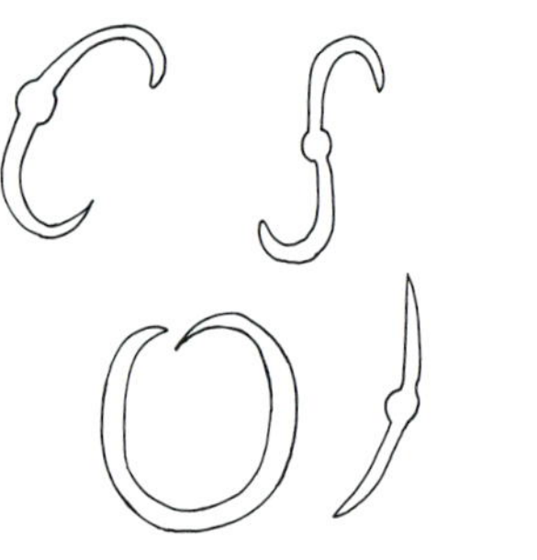

C-, O- and S-shaped rods

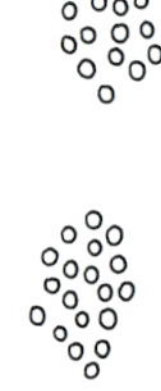

miliary granules of body wall

(after Cutress, 1996)

(after Sluiter, 1910)

EXPLOITATION:

Fisheries: *Astichopus multifidus* is harvested by SCUBA and hookah diving. There is an artisanal fishery in Panama, where this species is one of three most important species in the commercial catches, although in low numbers. In 1997, fishing in the Bocas del Toro region (Panama) seriously affected populations of *A. multifidus*. This species was absent in 95% of the protected areas around Cayo Zapatillas (Panama), which suggests that the local peoples may have overexploited this species from the marine park. It is commonly part of multispecies fisheries that include *I. badionotus* and *H. mexicana*.

Regulations: There is a complete fishing ban in Panama as of 2003.

Human consumption: Mostly, the reconstituted body wall (bêche-de-mer) is consumed by Asians.

Main market and value: The main market is China. Market value not determined.

LIVE (photo by: F. Charpin)

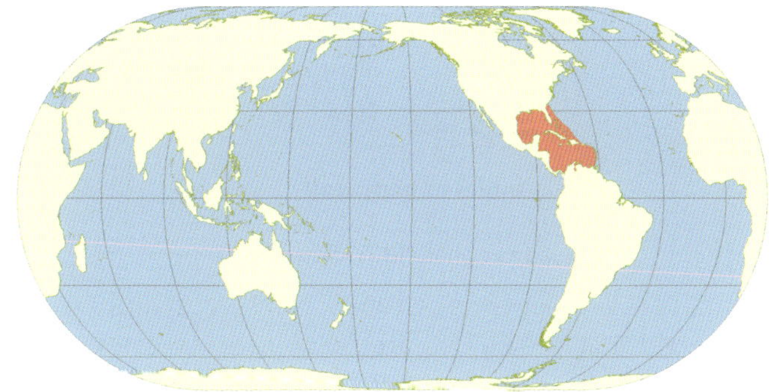

GEOGRAPHICAL DISTRIBUTION:
It can be found in the Caribbean Region, including Florida (USA), the Bahamas, Panama.

Australostichopus mollis (Hutton, 1872)

COMMON NAME: Brown mottled sea cucumber.

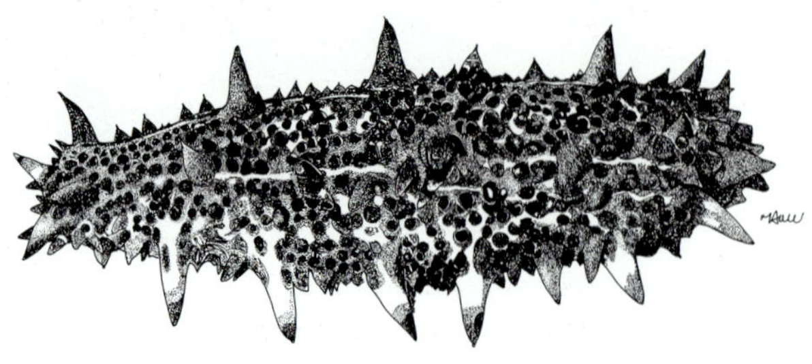

DIAGNOSTIC FEATURES: Colour may vary from blackish-brown to brown to yellow or cream. The ventral surface is lighter in colour. Numerous, long and short, conical papillae occur on the dorsal surface and along the lateral margins of the ventral surface, which may be lighter in colour (e.g. yellowish) or dark brown. Mouth is ventral with 20 relatively small, near-black tentacles. Anus is terminal.

Ossicles: Tentacles with large (about 800 µm long) curved and spiny rods and small, up to 150 µm long, and smooth ossicles of different shapes. Dorsal body wall with tables with round discs, 90–100 µm across, perforated by 4 central and 4 peripheral holes (occasionally a whole circle of perforations), with a spire ending in a Maltese cross. Ventral body wall with similar tables, but slightly smaller, that have discs up to 70 µm across and rarely rosettes, about 90 µm long. Ventral podia with perforated plates.

Processed appearance: Light brown to greyish-brown and with a mottled colour pattern. Papillae on the dorsal surface should be evident, but much reduced when compared to the live animals.

Remarks: Previously known as *Stichopus mollis*. The designation to the genus *Australostichopus* was based on morphological traits, such as the possession of only regular table ossicles in the body wall, and a glycoside molecule that is not found in other *Stichopus*.

Size: Average fresh weight about 110 g; average fresh length 17 cm.

HABITAT AND BIOLOGY:

This species is found in depth ranges of 5–100 m. Inshore, it occurs on sandy mud, sand and in intertidal rock pools. Offshore, it is found on muddy sand and on shell deposits. Inhabits bedrock and boulders. It is also regularly found at the bases of large kelp (e.g. *Ecklonia*).

It attains size-at-maturity at 75 g, it reproduces annually from October to February.

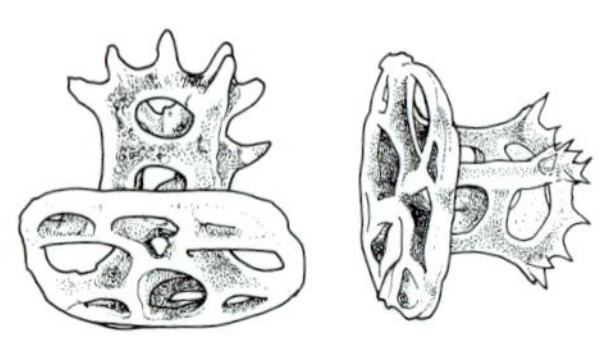

tables of body wall

(source: photo M.A. Sewell)

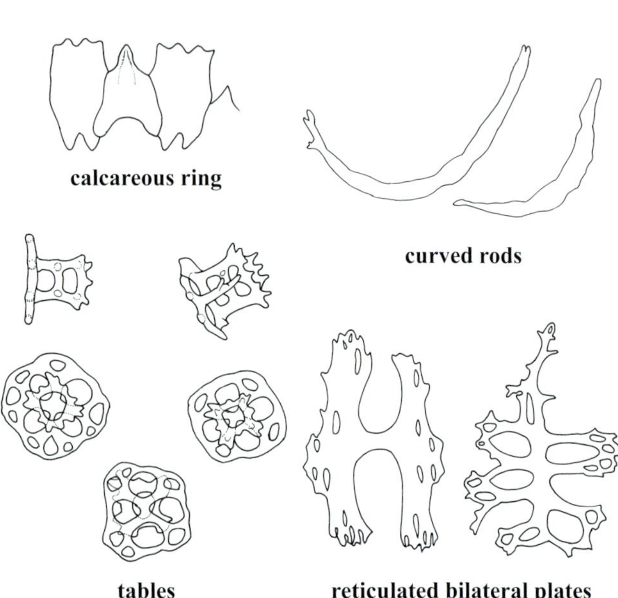

calcareous ring

curved rods

tables

reticulated bilateral plates

(after Dendy, 1897)

EXPLOITATION:

Fisheries: Semi-industrial. It is currently fished only in New Zealand by hand-gathering while free-diving; or as a bycatch from dredge fisheries for scallops.

Regulations: The fishery for this species in New Zealand is managed by means of a TAC, no-take reserves, and a quota management system and gear restrictions.

Human consumption: Mostly, the reconstituted body wall (bêche-de-mer) is consumed by Asians.

Main market and value: In New Zealand, there is a small fishery for this species (about 6 tonnes/year) and the average price achieved at first sale is about USD275 kg^{-1} dried.

LIVE (photo by: K. Clemments)

PROCESSED (photo by: L. Zamora)

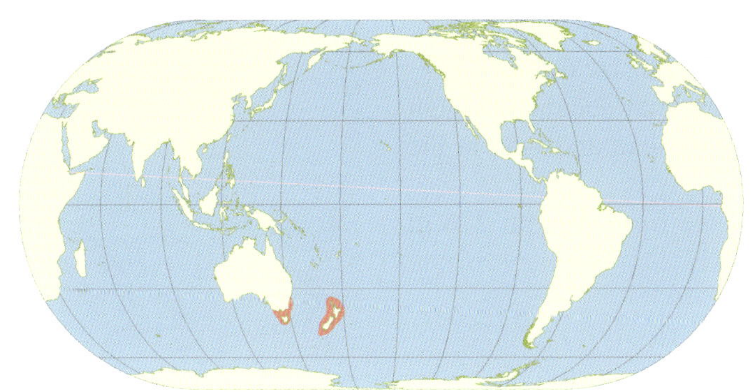

GEOGRAPHICAL DISTRIBUTION:

Found throughout New Zealand (including Snares Islands to south), and in south-eastern parts of Australia (south coast of New South Wales, Victoria and Tasmania).

Isostichopus badionotus (Selenka, 1867)

COMMON NAMES: Sea cucumber, Four-sided sea cucumber (**FAO**), Chocolate chip cucumber (Colombia), Pepino de mar (Panama), Sea pudding.

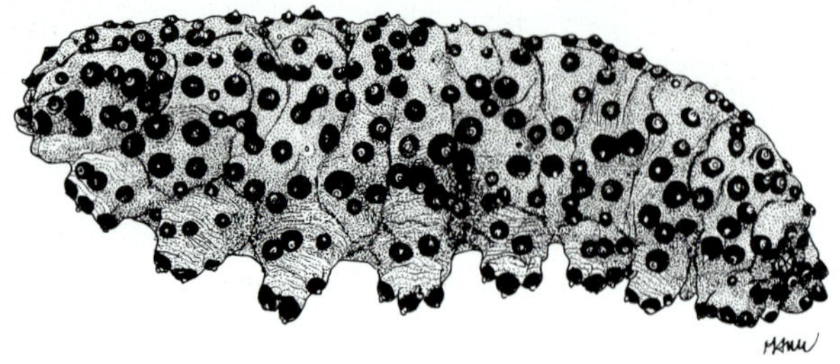

DIAGNOSTIC FEATURES: Brown, beige, yellow or black dorsally. The body is relatively firm and elongated with rounded anterior and posterior ends. Body shape is subcylindrical; arched dorsally and flattened ventrally. The dorsal surface is covered with brownish-black blunt papillae. The papillae at the lower lateral margins are long, robust and conical shaped with rounded tips. The mouth is ventral with 20 large tentacles. The anus is terminal.

Ossicles: Tentacles of largest specimens with spiny rods, 50–1100 μm long, and tables with discs 65–100 μm across as well as some C-shaped rods, 70 μm long on average. Body wall with numerous tables and C-shaped rods. Table discs regular, smooth, 40–60 μm across, perforated by 4 central holes and 1 complete circle of 10–12 peripheral holes; spire ending in a spiny crown. C-shaped ossicles, 50–70 μm long.

Processed appearance: Cylindrical and moderately elongated with rounded ends. Small wrinkles are evident on the dorsal surface. Dark brown dorsally; lighter coloured ventrally. The dorsal surface is rough and covered with dark spots, while the ventral surface is grainy. Common size 6–12 cm.

Remarks: Studies have shown that this species may accumulate high levels of trace metals including copper, nickel, lead and zinc in the digestive tract.

Size: Maximum length about 45 cm. Average fresh weight: 276 g; average fresh length about 21 cm (Cuba).

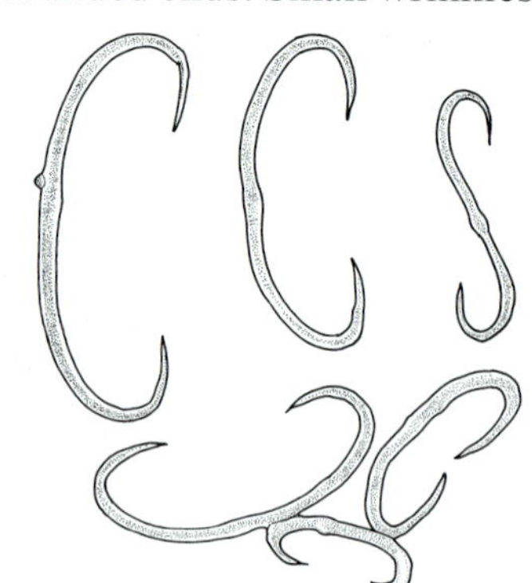

C-shaped elements of body wall

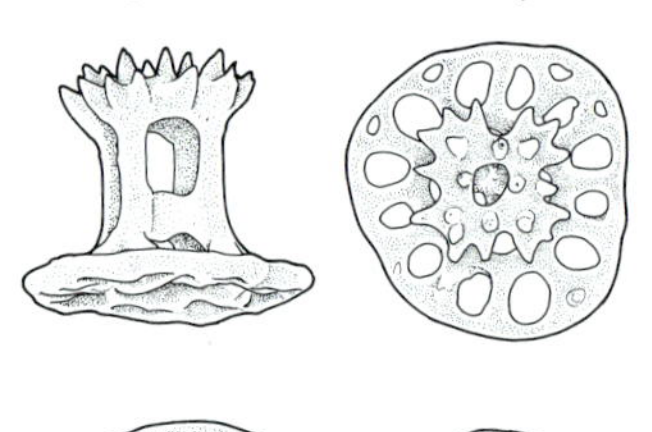

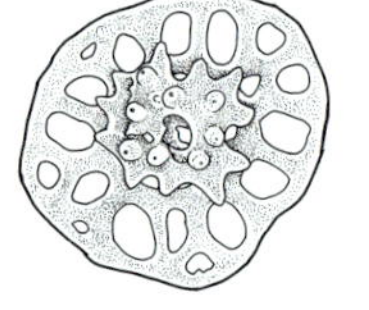

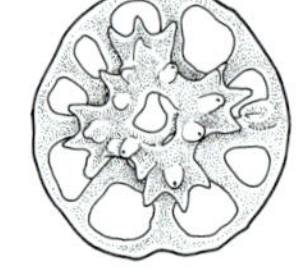

tables of body wall

HABITAT AND BIOLOGY: In Cuba, this species prefers sandy-muddy or sandy habitats, interspersed with seagrass or algae. Found between 0.5 and 19 m deep. In Colombia, *Isostichopus badionotus* prefers coral reefs, seagrass beds, rubble bottoms and sandy bottoms. Adults are generally non-cryptic, whereas the juveniles are reported to hide among coral rubble. It is a sluggish species that occurs on soft, shaded bottoms in shallow water, and shows no tendency toward concealment.

It attains size-at-maturity between 170 and 250 g. In Venezuela (Bolivarian Republic of), half of the animals reach sexual maturity at 30 cm and has a continuous reproductive season. In Panama, this species reaches sexual maturity between 13 and 20 cm, and has a peak in reproductive activity between July and November. In Brazil, spawning seems to occur from October to February, with a peak in January when seawater is warm.

(source: photo Giomar Borrero-Perez)

EXPLOITATION:

Fisheries: In the Caribbean, it is one of the most important commercial species. This species is collected in artisanal and semi-industrial fisheries. It is fished commercially in Cuba, Nicaragua and Venezuela (Bolivarian Republic of). This has been the only species fished in Cuba since 1999 despite the availabity of other species. In Colombia, there is an illegal, unregulated and non-quantified fishery for this species, and it is of potential commercial interest in Florida (USA), Puerto Rico and the United States Virgin Islands.

Regulations: In Cuba, the fishery is managed through a minimum legal length of 24 cm (or 22 cm ventrally), a fishing season between 1 June and 31 October, it is only open to artisanal fishers and there are no-take reserves. In Cuba, landings are closely monitored and compared with data on sale and exports; there is only one export company and logbooks must be submitted prior to shipments.

Human consumption: Mostly, the reconstituted body wall (bêche-de-mer) is consumed by Asians. It is occasionally consumed locally for medicinal purposes.

Main market and value: Hong Kong China SAR. It is sold by fishers in the Caribbean at USD22 kg^{-1} salted. Some salted product is processed to dried form in Chinese processing plants. Prices in Hong Kong China SAR retail markets ranged from USD203 to 402 kg^{-1} dried.

LIVE (photo by: E. Ortiz)

PROCESSED (photo by: S.W. Purcell)

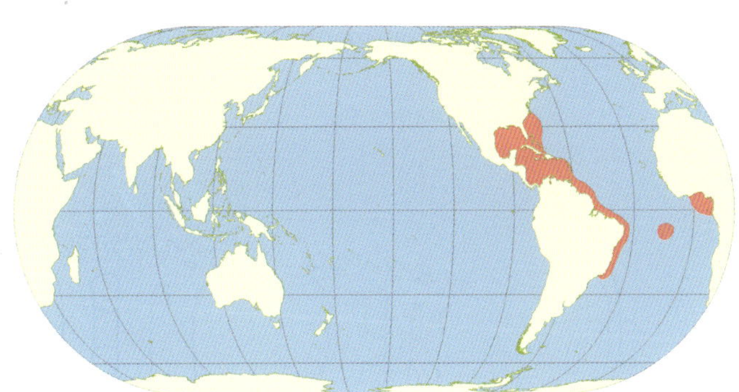

GEOGRAPHICAL DISTRIBUTION:

Widely distributed throughout the Caribbean Sea, from subtropical Atlantic, Brazil, Venezuela (Bolivarian Republic of), Colombia, Panama, Yucatan (Mexico) to southern Florida and the Bahamas also in South Carolina (USA), the Mid Atlantic at Ascencion Island, in the Gulf of Guinea off Western Africa.

COMMON NAMES: Brown sea cucumber, Giant sea cucumber (**FAO**), Pepino de mar gigante (**FAO**), Concombre de mer géant (**FAO**), Pepino de mar.

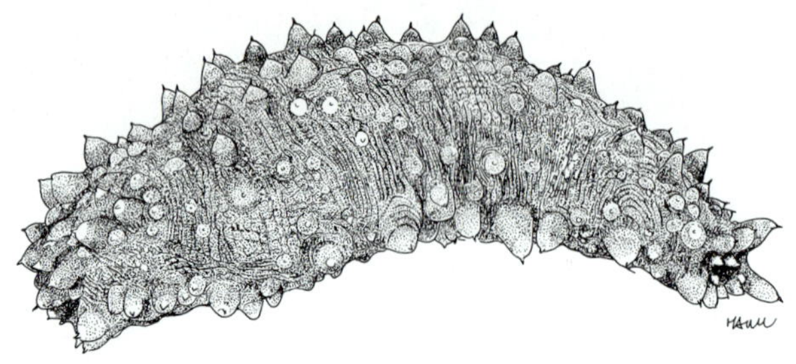

DIAGNOSTIC FEATURES:
Dorsal surface is dark brown with numerous stout, pointed, yellow papillae. Dorsal papillae are not arranged in rows. The ventral podia are arranged in a double row in the mid and in single rows along the sides. The body is subcylindrical; arched dorsally and flattened ventrally. Ventral surface lighter brown with podia in dense bands. The mouth is ventral with 20 yellow tentacles with markedly wide disc. Anus is terminal with no teeth.

Ossicles: Tentacles with curved rods, 100–400 µm long. Body wall with tables and C-shaped rods; rosettes and X-shaped rods absent. Table discs, on average 40 µm across, smooth, and with a moderately high spire that ends in a spiny crown. C-shaped rods 40 µm long.

Processed appearance: Dried *Isostichopus fuscus* are relatively stout with rounded, conical, papillae dispersed over the entire body. The papillae are especially numerous at the lower lateral margins. Coloration is greyish black. Common dried size 6–10 cm.

Size: Average fresh weight from 294 g (Ecuador) to 497 g (Mexico); average fresh length from 20 cm (Ecuador) to 24 cm (Mexico).

HABITAT AND BIOLOGY: *I. fuscus* can be found on rocky and coraline patches on the coastal zone from the shallow subtidal to 39 m depth. In Galapagos, it prefers rocky bottoms where the green algae *Ulva* sp. is predominant. In Baja California, it is found in coral and rocky habitats.
In the Galapagos Islands, this species attains size-at-maturity at 160 to 170 g drained weight, and has a continuous reproduction through the year. In Mexico, it attains size-at-maturity at 367 g (20 cm) and has an annual reproductive event during summer.

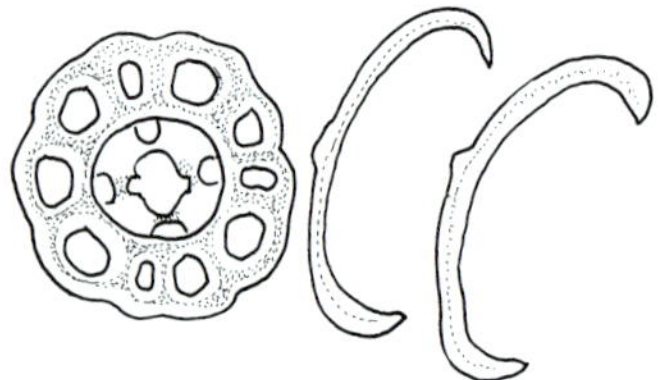

table disc and C-shaped rods

(source: photo C. Hickman)

C-shaped rod

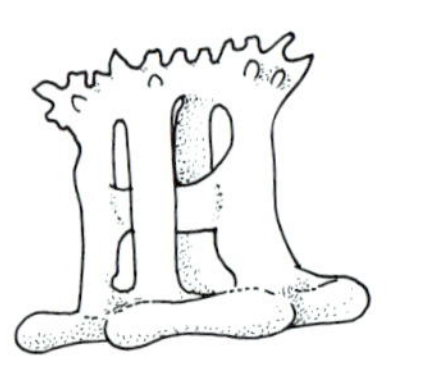

tables

(after Deichman, 1958)

EXPLOITATION:

Fisheries: *I. fuscus* is harvested in semi-industrial fisheries by hookah diving. This species is under commercial exploitation in Ecuador, Mexico, Panama and Peru. In Panama, it is fished illegally. In the Galapagos Islands, there was a moratorium on fishing in 2009 and 2010 as the minimum population density required to open the fishery (11 ind. 100 m^{-2}) was not met. At mainland Ecuador, the fishery started in 1988 and fishers serially deplete the fishing grounds. The fishery in Mexico, Central and South America started after the depletion of sea cucumbers in traditional fishing grounds.

Regulations: In the Galapagos Islands, this fishery is managed by means of a TAC, minimum legal length (20 cm fresh or 7 cm dry), no-take reserves, a fishing season (two months), and access is only to artisanal fishers that are permanent residents of the islands. In Mexico, there is a fishing season (October–May), a minimum legal size (400 g or 20 cm) and annual permits. Additionally, no-take reserves are established where the smallest individuals are found. There is a total ban on fishing *I. fuscus* in continental Ecuador. In Mexico, adaptive management includes quotas, catch reports and stock monitoring. This species is the only commercially exploited sea cucumber, so far, that is listed in CITES Appendix III.

Human consumption: The reconstituted body wall (bêche-de-mer) is consumed by Asians.

Main market and value: Hong Kong China SAR and the United States of America. It is sold at USD1.4 per unit fresh.

LIVE (photo by: S.W. Purcell)

PROCESSED (photo by: S.W. Purcell)

GEOGRAPHICAL DISTRIBUTION:

Found from Baja California to mainland Ecuador, including Galapagos, Socorro Island, Cocos (Keeling) Islands, Malpelo and Revillagigedos islands. Hooker, Solís-Marín and Leellish (2005) include Peru (Islas de Lobos de Afuera) in its geographical distribution.

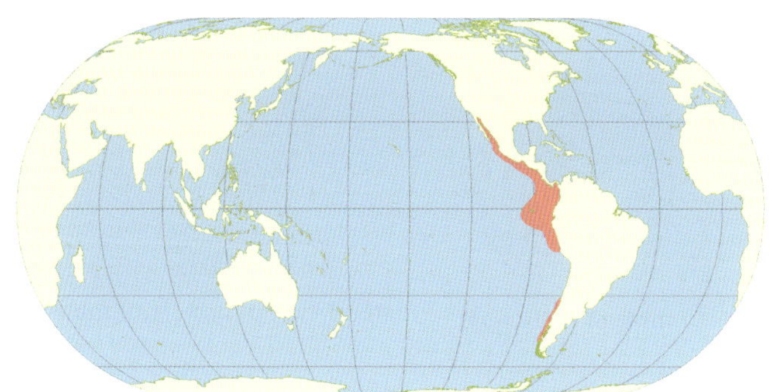

Parastichopus californicus (Stimpson, 1857)

COMMON NAMES: Giant red sea cucumber (**FAO**), Giant California sea cucumber.

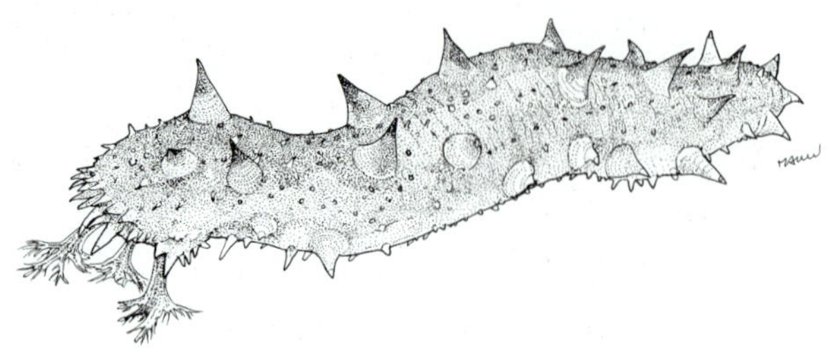

DIAGNOSTIC FEATURES:

Mottled brown to red and yellow over the dorsal surface. Juveniles tend to be solid red or brown, without mottling. Body is cylindrical and quite elongated, with slightly tapered ends. Dorsal surface with both large (about 40 in total) and small fleshy papillae, which are yellow to orange in colour and red tipped. Ventral surface light cream in colour. Podia are numerous ventrally, arranged in 5 rows. Mouth ventral with 20 short tentacles.

Ossicles: Body wall with tables and buttons. Tables with round discs that are 70–95 µm across, perforated by 4 central holes with which 4 smaller and more distal holes alternate, but often there are smaller holes on each side of the latter; high spire ending in a spiny crown. The irregular buttons, some 90 µm long, have 5–7 pairs of holes.

Processed appearance: Processed animals may be 10–30 cm long. Dried animals are grey, and the long papillae are still evident. A cut is normally made on the ventral surface.

Remarks: This species is somewhat similar in appearance to *Apostichopus parvimensis*, which in contrast, has a brownish colour dorsally and is lighter in colour ventrally, and has more numerous small black-tipped papillae.

Size: Maximum length about 50 cm. Average adult fresh weight at least 500 g; average length from 25 to 40 cm.

HABITAT AND BIOLOGY: It occurs on a wide variety of substrates and current regimes, in water depths from the intertidal to 250 m. This species is most abundant in areas of moderate current on cobbles, boulders or crevassed bedrock, but avoids muddy bottoms and areas with freshwater runoff.

It reproduces annually during the summer dry season (May–August) and reproduction seems to be correlated with bright sunshine days and high phytoplankton productivity.

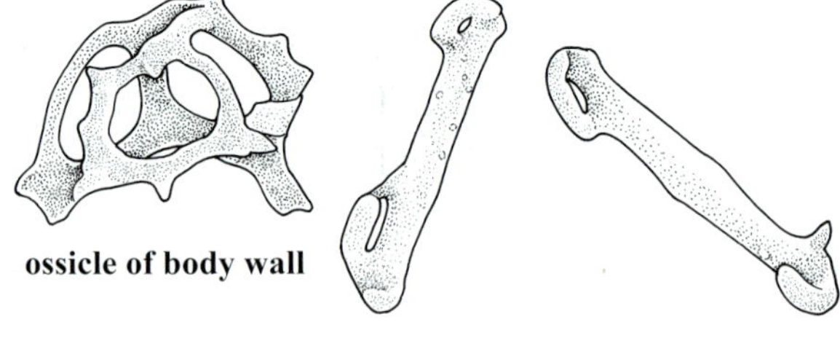

ossicle of body wall

rods of body wall

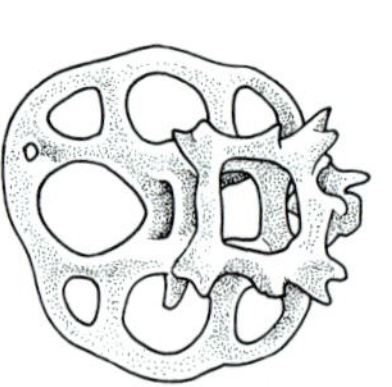

table of body wall

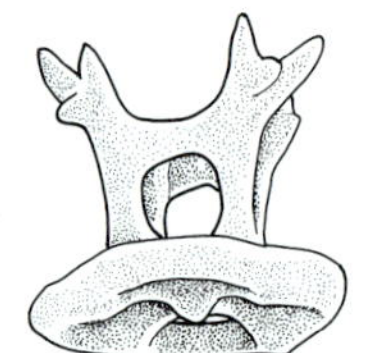

lateral view of body wall table

(source: Solís-Marín *et al.*, 2009)

EXPLOITATION:

Fisheries: *Parastichopus californicus* is exploited in industrial fisheries by hand collection by SCUBA diving and by trawling. The fishery of this species in Washington (USA) started in the 1970s, later spreading to California, Alaska and Oregon. In British Columbia (Canada), the first official landings date back to 1980. It is the only commercially harvested species on the west coast of Canada (Vancouver Island, Georgia and Johnstone Straits and Pudget Sound). Until 1997, an average of 75% of the annual catch was from the southern California trawl fishery. In Alaska (USA), *P. californicus* is one of the main species harvested for subsistence in native communities.

Regulations: In Canada, the fishery is managed by means of a fishing season (October and November), limited number of licences, no-take marine reserves, TACs, and individual transferable quotas (ITQs). In California, exploitation of this species is regulated by permits for each gear type and limited entry restrictions, but there is no TAC regulation. Trawling is prohibited in some conservation areas. There are no-take marine reserves to protect breeding populations.

Human consumption: Mostly, the reconstituted body wall (bêche-de-mer) is consumed by Asians. The muscle strips of this species are also exported and consumed.

Main market and value: The main markets are Hong Kong China SAR, Taiwan Province of China, Mainland China and the Republic of Korea. It is sold by fishers for up to about USD3.70 kg^{-1} wet (gutted).

LIVE (photo by: J.M. Watanabe)

PROCESSED (photo by: J. Akamine)

GEOGRAPHICAL DISTRIBUTION:
Distributed along the Pacific coast of North America, from the Aleutians Islands, Alaska to the Gulf of California.

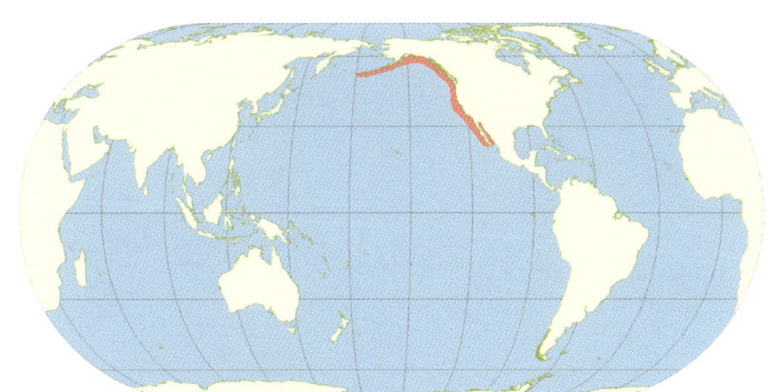

Stichopus chloronotus Brandt, 1835

COMMON NAMES: Greenfish (**FAO**), Trepang vert (**FAO**), Zanga sono, Mahitsikely (Madagascar), Barbara (Mauritius), Mullu attai, Pavaka attai (India), Saâu bieǎn, Ñoät beâ ô (south Viet Nam), Cuatro cantos (Philippines), L'ananas vert (New Caledonia), Tekirin (Kiribati), Holomumu (Tonga), Dri-votovoto (Fiji).

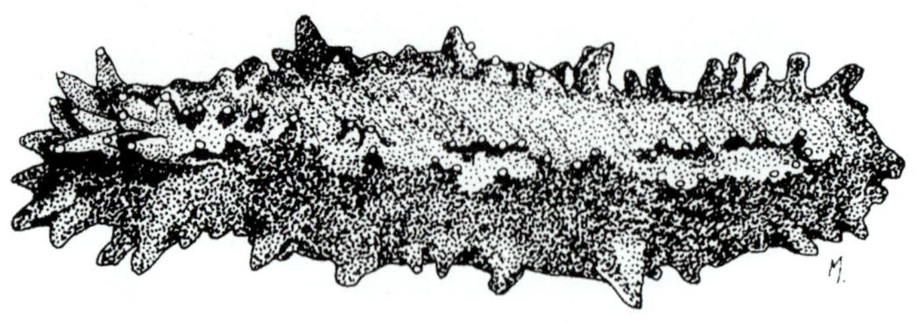

DIAGNOSTIC FEATURES: Body colour is dark green to near black dorsally; dark green ventrally. Rows of long, conical papillae on both sides of the dorsal surface and along both lower lateral margins of the body. Papillae tips are usually, but not always, orange to yellow. Body moderately firm and squarish in cross-section. Ventral podia are long and green, in four rows. Mouth is ventral with 19 or 20, white to greyish, stout tentacles. Anus is terminal bordered by five large papillae.

Ossicles: Base of tentacles are tables reduced to the disc, small knobbed rods and few very long rods, up to 450 µm long. Tips of the tentacles have spiny curved rods, 65–470 µm long. Dorsal and ventral body wall with tables and C-shaped rods. Ventrally, tables are larger, with discs from 30–45 µm across, than dorsally, with discs 25–30 µm across. Table discs are smooth and perforated by 4 central and 4–10 peripheral holes with a pillar ending in a crown of spines that resembles a Maltese cross. C-shaped rods are small, up to 50 µm long. Dorsal papillae have characteristic larger tables, with discs 55–80 µm across, large C-shaped rods, up to 70 µm long, and irregular rods. Ventral podia have reduced tables, few irregular C-shaped rods, 40–100 µm long, and rods of 270–470 µm long, and perforated plates.

Processed appearance: Moderately elongate with a squarish cross-section; each of the four edges of the body is covered with pointy wart-like projections. The entire body is dark grey to black. No cuts or small cut across mouth. Common dried size 10–12 cm.

Remarks: The body-wall may disintegrate if the animal is held out of the water for a long time.

Size: Maximum size 35 cm, mostly 20 cm when full grown. Average fresh weight: 80 g (Mauritius), 100 g (Papua New Guinea, India), 100–400 g (Réunion), 150 g (New Caledonia); average fresh length: 8–20 cm.

HABITAT AND BIOLOGY: An inhabitant of coral reefs, in shallow waters from the intertidal to depths of 10 m. *Stichopus chloronotus* can be found on reef-flats and upper reef slopes

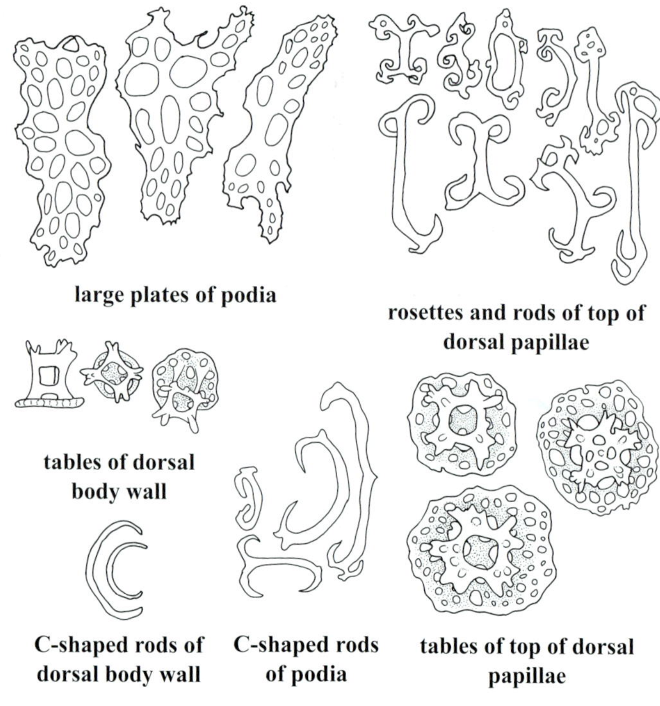

large plates of podia

rosettes and rods of top of dorsal papillae

tables of dorsal body wall

C-shaped rods of dorsal body wall

C-shaped rods of podia

tables of top of dorsal papillae

(after Massin *et al.*, 2002)

at densities up to 1 ind. m^{-2} in shallow areas. Found mostly on coarse coral sand and sheltered habitats with coral rubble. In China, it is reported in deeper water (40 to 60 m). Sexual reproduction is bi-annual during summer. It undergoes asexual reproduction (fission) mostly during the cool season.

EXPLOITATION:

Fisheries: Exploited in artisanal (e.g. Fiji, Tonga, Viet Nam) and semi-industrial fisheries (Mauritius). It is harvested by hand by gleaning at low tide or by free diving in many localities throughout the Indo-Pacific, by compressed-air diving and gleaning in Mauritius and Viet Nam. It is seldom harvested in fisheries of the Coral Sea and Great Barrier Reef (Australia). This species is harvested for subsistence consumption in Wallis and Futuna Islands and French Polynesia. In Asia, it is of commercial

LIVE (photo by: S.W. Purcell)

PROCESSED (photo by: S.W. Purcell)

importance in China, Japan, Malaysia, Thailand, Viet Nam, Indonesia (heavily exploited) and the Philippines. In Kenya and Madagascar, it is harvested but it is of low commercial value.

Regulations: Before a national fishing moratorium in Papua New Guinea, regulations included a minimum legal length of 20 cm live and 10 cm dry. In the Great Barrier Reef fishery, there is a minimum legal length of 20 cm live, there are permits and limited entry regulations, and this species is subject to a rotational harvest strategy and catches are limited in a combined TAC with some other species.

Human consumption: Mostly, the reconstituted body wall (bêche-de-mer) is consumed by Asians. In certain Pacific islands, it is eaten as part of traditional subsistence diets.

Main market and value: Main export markets are Singapore and Hong Kong China SAR. Traded recently at about USD60 kg^{-1} dried in the Philippines. Previously sold for USD12–17 kg^{-1} dried in Papua New Guinea, and for USD17–20 kg^{-1} in Viet Nam. In New Caledonia, it is exported for about USD25 kg^{-1} dried. In Fiji, fishers receive USD0.4–0.7 per piece fresh. Prices in Guangzhou wholesale markets ranged from USD63 to 95 kg^{-1} dried.

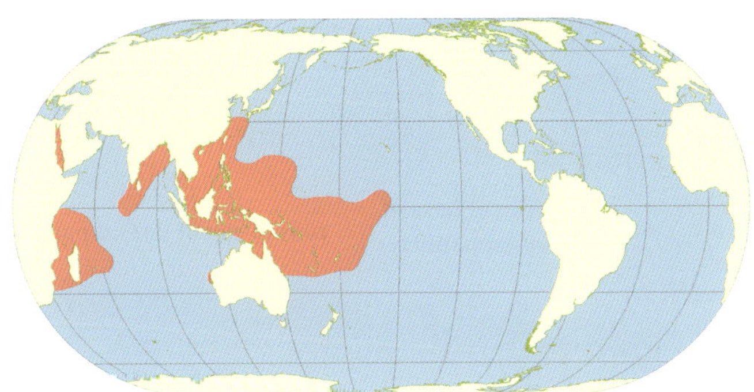

GEOGRAPHICAL DISTRIBUTION:
Islands of western Indian Ocean, Mascarene Islands, East Africa, Madagascar, Maldives, Sri Lanka, Bay of Bengal, East Indies, North Australia, the Philippines, China and southern Japan, most of the islands of the Central Western Pacific but apparently absent from the Marshall Islands.

Stichopus herrmanni Semper, 1868

COMMON NAMES: Curryfish, Mul attai (India), Marhm (Egypt), Ñoät ngaän ñaù, Ñoät ngaän tröôơng (south Viet Nam), Trakitera, Crampon (Madagascar), Tairi (Zanzibar, Tanzania), La curry (New Caledonia), Lomu (Tonga), Tekare (Kiribati), Laulevu (Fiji).

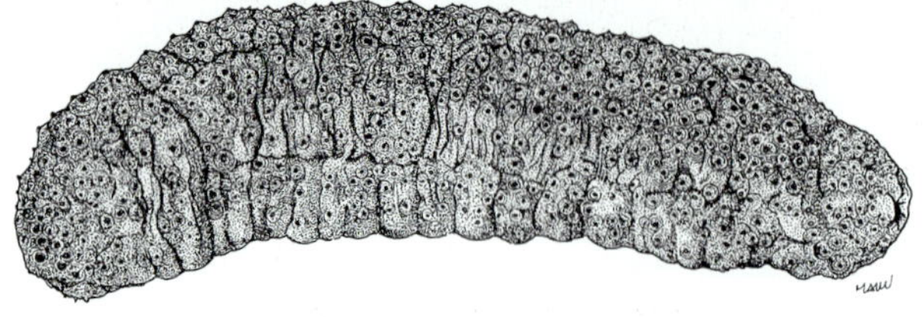

DIAGNOSTIC FEATURES: Body colour varies from light mustard-yellow to orangey-brown or brown or olive green. Colour tends to be lighter ventrally. Numerous dark brown to black spots scattered over the entire body; two double-rows of larger wart-like papillae, bordered by fine dark rings. Podia are numerous ventrally. Body relatively firm, moderately elongate and squarish in cross-section. Mouth is ventral with 8–16 stout green tentacles. Anus is terminal, with no teeth nor surrounding papillae.

Ossicles: Tentacles with spiny, slightly curved rods, some times forked and/or perforated distally, 60–850 µm long. Dorsal and ventral body wall with the same type of tables, C-shaped rods and rosettes. Table discs 25–45 µm across, perforated by 4 central and 4–8 peripheral holes, spire ending in a narrow, often spiny crown. Rosettes 25–55 µm long. C-shaped rods 35–100 µm long. Ventral podia have reduced tables, 30–45 µm across, large perforated plates with the median part often enlarged and perforated, and rods, 200–360 µm long. Dorsal papillae with rods up to 200 µm long, C- or S-shaped ossicles similar in size and shape as those of the body wall, rosettes and tables up to twice the size as those of the body wall.

Processed appearance: Relatively elongate and squarish in cross-section. Body tends to be various shades of beige to brown. Dorsal surface wrinkled and covered with small, dark, bumps, while the underside is smoother. No cuts or small cut across mouth. Common dried size 12 to 18 cm.

Remarks: Previously known as *Stichopus variegatus*. The body wall disintegrates if the animal is held out of water for a long time. *S. monotuberculatus* was also reported as *S. variegatus*, so the distributional range of *S. herrmanni* may be biased.

Size: Maximum size 55 cm, mostly 20–40 cm. Average fresh weight: 1 000 g (Papua New Guinea, Egypt, India), 1 100 g (Viet Nam), 1 680 g (New Caledonia) and 1 000–2 500 (Réunion).

HABITAT AND BIOLOGY: Occurs in a wide range of shallow tropical habitats. In the western central Pacific, *S. herrmanni* prefers seagrass beds, rubble and sandy-muddy bottoms between 0 and 25 m. In the Africa and Indian Ocean region, it can be found in lagoons, seagrass beds and rubble over sandy-muddy bottoms between 0 and 5 m. Juveniles found more commonly in shallow waters.

It attains size-at-maturity at about 31 cm and reproduces annually during summer. In Réunion, it reproduces during the dry season.

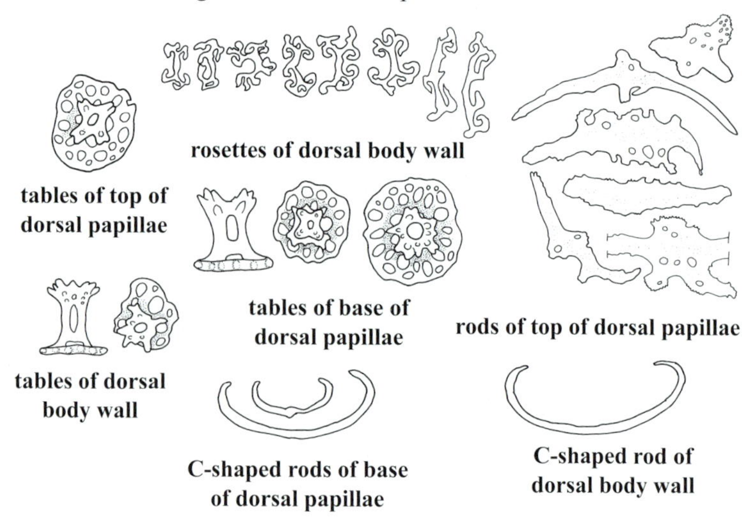

(after Massin *et al.*, 2002)

EXPLOITATION:

Fisheries: Mostly exploited artisanally. Collected by gleaners and breath-hold divers in most localities throughout its distribution in the Indo-Pacific. In Viet Nam, fishers may also use hookah diving, and in Egypt it was collected using SCUBA. On the Great Barrier Reef (Australia), it has been collected in minor quantities but there is interest to target this species. Collected for subsistence in Palau, Wallis and Futuna Islands and French Polynesia. In Asia, it is commercially exploited in China, Malaysia, Thailand, Indonesia (heavily fished), the Philippines and Viet Nam. In Malaysia, this species is used commercially for the preparation of traditional medicinal products.

Regulations: Before the moratorium in Papua New Guinea, regulations included a minimum landing sizes (25 cm live, 10 cm dry). In New Caledonia, there is a minimum legal length of 35 cm for fresh animals and 15 cm when dried. On the Great Barrier Reef (Australia), there is a minimum landing size of 35 cm fresh, no-take reserves, strict licensing, and this species is subject to a rotational harvest strategy.

Human consumption: Mostly, the reconstituted body wall (bêche-de-mer) is consumed by Asians. In some Pacific islands, it is consumed either whole or its intestine and/or gonads as delicacies or as protein in traditional diets or in times of hardship (i.e. following cyclones). In Egypt, this species is used for the preparation of traditional medicinal products.

Main market and value: Singapore, Hong Kong China SAR, Egypt and Korea. It has been traded recently at USD35–58 kg^{-1} dried in the Philippines. It is sold for USD20 kg^{-1} dried in Viet Nam, and was previously sold for USD12 kg^{-1} dried in Papua New Guinea. In New Caledonia, *S. herrmanni* is exported for USD20–30 kg^{-1} dried. In Fiji, fishers receive USD1–5 per piece fresh. Prices in Hong Kong China SAR retail markets ranged from USD182 to 214 kg^{-1}. Prices in Guangzhou wholesale markets ranged from USD79 to 159 kg^{-1} dried.

LIVE (photo by: S.W. Purcell)

PROCESSED (photo by: S.W. Purcell)

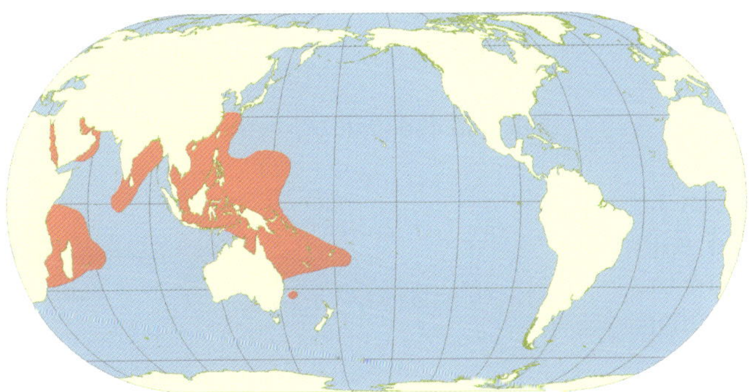

GEOGRAPHICAL DISTRIBUTION:

Mascarene Islands, East Africa and Madagascar, Red Sea, southeast Arabia, Gulf of Aqaba, Persian Gulf, Maldives, Sri Lanka, Bay of Bengal, East Indies, North Australia, the Philippines, China and southern Japan. It occurs in most countries of the western Pacific as far east as about Tonga and as far south as Lord Howe Island.

COMMON NAMES: Selenka's sea cucumber (**FAO**), Warty sea cucumber (Ecuador), Dragonfish (India, Papua New Guinea), Pepino de mar (Ecuador and Galapagos Islands), Mul attai (India), Hanginan and Loaf bread (Philippines), Smurf (Madagascar), Sankude (Zanzibar, Tanzania).

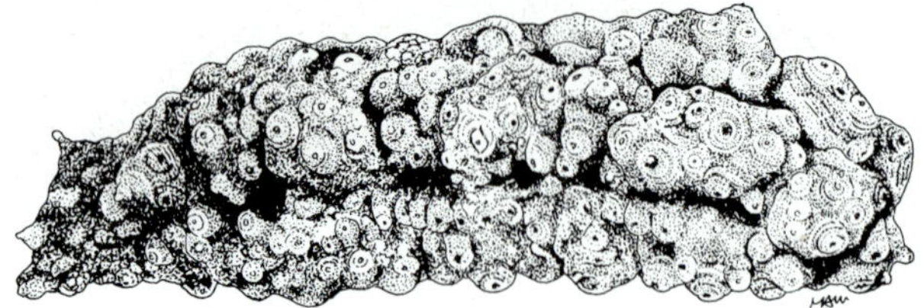

DIAGNOSTIC FEATURES: Coloration is highly variable, from grey to beige to dark red, dark brown or black with different coloured blotches dorsally. In India, it may be greenish-brown. Dorsal surface lightly arched with long and conical, or wart-like, papillae mostly in two rows along the upper dorsal surface and a row of larger papillae along the lateral margins of the flattened ventral surface. This species is relatively small. Numerous, large podia occur on the ventral surface. The mouth is ventral with 20 tentacles.

Ossicles: Tentacles with straight to nearly U-shaped rods, 60–700 µm long, some with forked extremities. Dorsal body wall with numerous tables, discs 25–35 µm across, and with a spire ending in a Maltese cross; numerous rosettes, 15–30 µm long; and few C-shaped rods, 45–60 µm long. Ventral body wall with tables, discs 30–55 µm across, and few C-shaped rods, 55–85 µm long. Ventral podia with tables 40–120 µm across, large perforated plates, and rods, 390–500 µm across, some with huge central perforated process. Dorsal papillae at their base with similar ossicles as those of the body wall, but with at their top huge tack like tables, 130–155 µm across, C-shaped rods, 45–80 µm long, and rods with a large central perforated process.

Processed appearance: Dried animals are slender with a squarish cross-section; the dorsal surface retains the wart-like bumps and the ventral surface is smoother. In Ecuador, dried *Stichopus horrens* are smoky black with spiky papillae. Ventral surface flattened with podia visible. In Papua New Guinea, the dried animals tend to be brown to brownish-black. No cuts or small cut across mouth. Common dried size 8–12 cm.

Remarks: It can be mistaken for *S. monotuberculatus*, *S. naso* or *S. quadrifasciatus*. One distinguishing feature from other species is the huge 'thumb-tack' shaped table ossicles of the dorsal papillae. Certain sightings of this species may be other species, so distribution records may be in error.

Size: Average fresh weight from 110 g (Philippines) to 200 g (India, Papua New Guinea); average fresh length 12 cm (Philippines), 20 cm (India, Papua New Guinea) and 23 cm (Ecuador).

HABITAT AND BIOLOGY: This species can be found mostly on rocky bottoms interspersed with sandy patches, between 2 and 20 m depth. In the western central Pacific, it can be found in reef flats and upper slopes. In East Africa and in the Indian Ocean, it appears to prefer lagoons and seagrass beds over sand and rubble between 0 and 5 m deep.

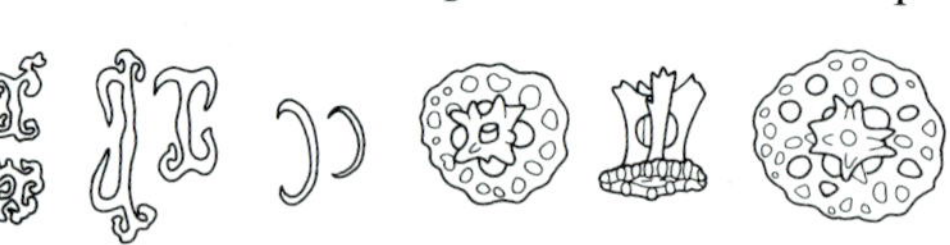

rosettes, C-shaped rods and tables of body wall

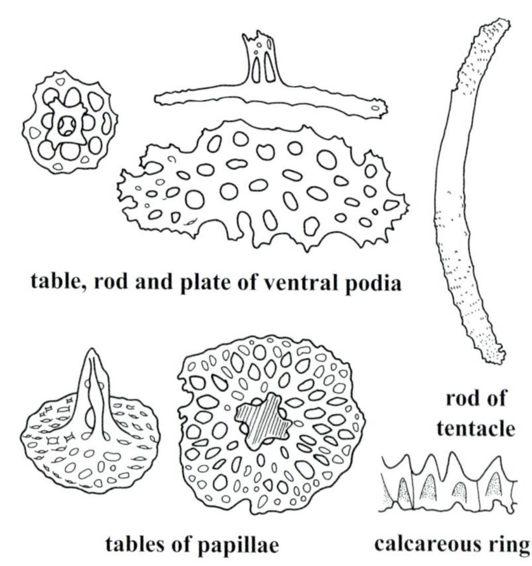

table, rod and plate of ventral podia

rod of tentacle

tables of papillae

calcareous ring

(after Cherbonnier, 1980)

In Madagascar, it can be found abundant on inner reef slopes, but also in seagrass beds, microatolls, and the detrital fringe and outer reef flat. In the Galapagos, it prefers rocky substrates between 5 and 20 m depth. This species is usually active nocturnally; during the day, it remains hidden in reef crevices, and its body may be smaller and wartless.

It attains size-at-maturity at 16–18 cm. It undergoes asexual reproduction by fission in Japan.

LIVE (photo by: G. Paulay)

EXPLOITATION:

Fisheries: This species is exploited in semi-industrial (Ecuador) and artisanal fisheries (e.g. Philippines, Malaysia). In the western Pacific region, it has been fished in many localities. In Asia, it is of commercial importance in China, Malaysia, Indonesia (heavily exploited) and the Philippines (one of the most sought-after species). In the

PROCESSED (photo by: S.W. Purcell)

Philippines, it is consumed by Muslims during the Ramadan season. This species is heavily exploited in Madagascar and collected illegally in the Galapagos Islands.

Regulations: In Moreton Bay (Australia), a mostly inactive fishery, there is a minimum legal length of 17 cm live.

Human consumption: Mostly, the reconstituted body wall (bêche-de-mer) is consumed by Asians. Its intestine and/or gonads are consumed in traditional diets. In Malaysia, it is used commercially for the preparation of traditional medicinal products and for the medicinal properties of its coelomic fluid or "gamat" and with medicinal purposes in China. These raw products are traditionally processed into gamat oil and gamat water, and recently into medicated balm, toothpaste and soap.

Main market and value: Hong Kong China SAR, China. In the Philippines, it is sold at USD39 kg^{-1} dried. Prices in Guangzhou wholesale markets ranged from USD56 to 83 kg^{-1} dried.

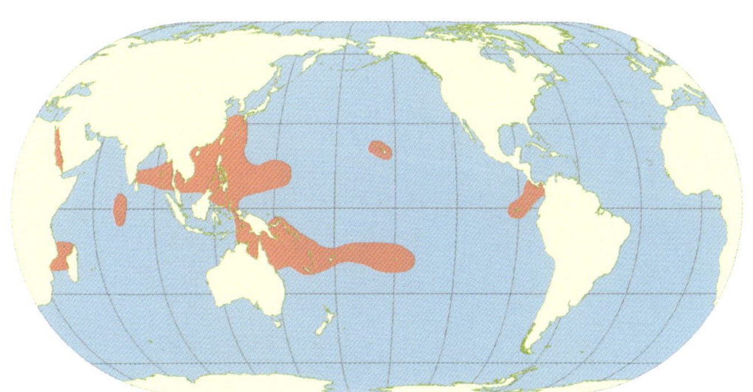

GEOGRAPHICAL DISTRIBUTION:

Red Sea, East Africa, Maldives, East Indies, North Australia, the Philippines, China and southern Japan, and islands of the Pacific, including the Galapagos Islands (Ecuador) and Hawaii (USA).

COMMON NAMES: Because *Stichopus monotuberculatus* is often confused with *S. horrens*, it has often been traded under the same common name of the latter.

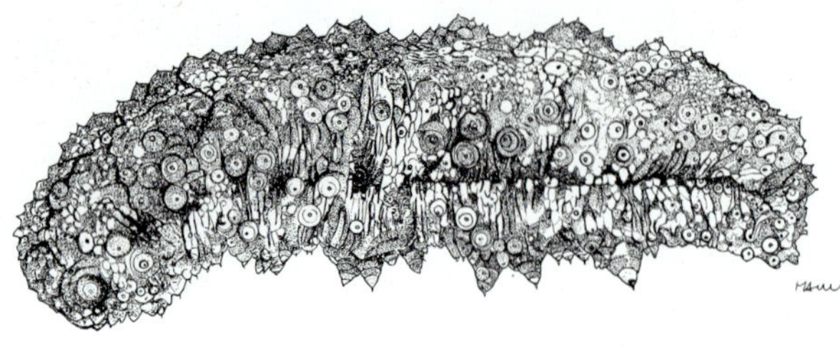

DIAGNOSTIC FEATURES: Coloration is highly variable, often indistinguishable from *S. horrens*: a grey to beige to yellowish background color dorsally overlaid with darker brown, grey to black spots and blotches, which can appear to form two indistinct transverse bands. Large pointed dorsolateral and ventrolateral papillae are retracted during the day. Trapezoidal to rectangular in cross-section. Ventral surface mottled similar to dorsal surface, with three longitudinal rows of large podia. The mouth is ventral with 20 tentacles. Byrne, Rowe and Uthicke (2010) state that *S. monotuberculatus* from tropical northeast Australia have low wart-like papillae and prominent lateral papillae and that specimens from the central Pacific had longer dorsal papillae.

Ossicles: Tentacles with rods of different sizes, very spiny at ends, 145–645 µm long. Dorsal body wall with tables, rosettes and C-shaped ossicles: tables 30–50 µm across, rim of discs is smooth, perforated by 4 central holes and 3–6 peripheral holes; spire ending in a wide spiny crown; rosettes some 20 µm long. Ventral body wall with tables of similar shape and size as those of the body wall and C-shaped rods, 60–70 µm long. Ventral podia with spiny rods, 250–415 µm long, with enlarged median process, unevenly perforated; spiny plates, 85–100 µm long; and tables with rounded but spiny disc. Dorsal papillae with tables, 45–70 µm across, and rods of various shapes and size; the largest ones are 135–350 µm long, have an enlarged median process.

Processed appearance: Probably the same as *S. horrens*. Dried animals are slender with a squarish cross-section; the dorsal surface retains the wart-like bumps and the ventral surface is smoother. Ventral surface flattened with podia visible. No cut. Common dried size 9–12 cm. In Papua New Guinea, the dried animals tended to be brown to brownish-black. No cuts or small cut across mouth. Common size: 8–12 cm.

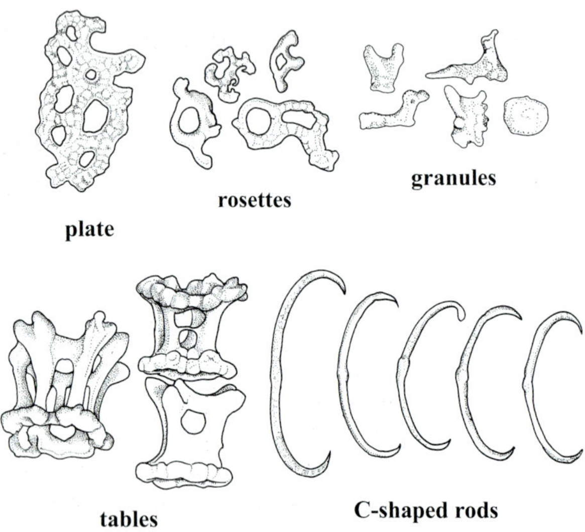

plate

rosettes

granules

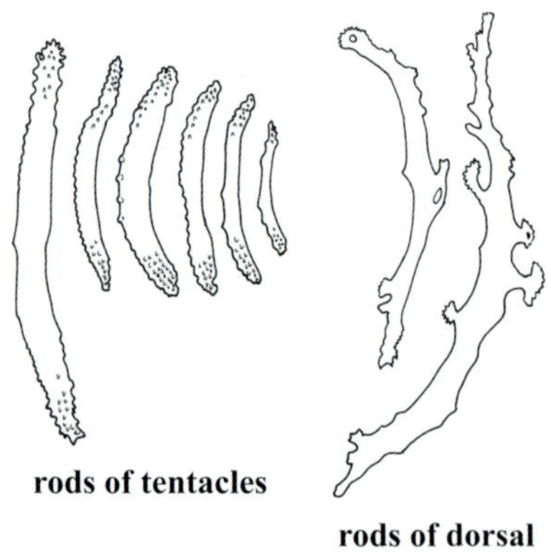

rods of tentacles

rods of dorsal papillae

tables

C-shaped rods

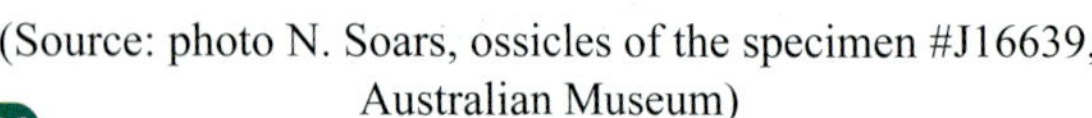

(Source: photo N. Soars, ossicles of the specimen #J16639, Australian Museum)

(after Massin, 1996)

Remarks: Note that this is not the *S. monotuberculatus* of Quoy and Gaimard (1833), which appears to be a western Indian Ocean endemic, but a widespread species that may be new. Often mistaken for *S. horrens*, *S. naso* or *S. quadrifasciatus*. This species lacks the huge 'thumb-tack' shaped table ossicles of the dorsal papillae that are present in *S. horrens* and *S. naso*. The genetic and morphological boundaries of these four *Stichopus* are currently uncertain and likely to be redefined.

Size: Probably the same as *S. horrens* where they both occur: average fresh weight from 110 g (Philippines) to 200 g (India, Papua New Guinea); average fresh length 12 cm (Philippines), 20 cm (India, Papua New Guinea).

LIVE (photo by: S.W. Purcell)

PROCESSED (photo by: S.W. Purcell)

HABITAT AND BIOLOGY: It hides during the day in crevices and under rubble, emerging at night on the reef flat, lagoons and reef slope from 1 m to at least 30 m depth. Reproduces via binary fission, at least on the Great Barrier Reef, Australia.

EXPLOITATION:

Fisheries: As this species is seldom recognized, the extent of exploitation is uncertain. It is probably exploited in artisanal fisheries in the Philippines and Malaysia, and for subsistence use in some Pacific Island countries.

Regulations: Because of its similarity in appearance to *S. horrens*, this species is inadvertently regulated along with the latter species.

Human consumption: Mostly, the reconstituted body wall (bêche-de-mer) is consumed by Asians. Its intestine and/or gonads are consumed in traditional diets; local consumption is the same as for *S. horrens*.

Main market and value: It is apparently sold along with *S. horrens* in Hong Kong China SAR, China and the Philippines, where the latter species is sold at USD39 kg^{-1} dried. In Fiji, it is exported for USD11–16 kg^{-1} dried. Prices in Guangzhou wholesale markets ranged from USD111 to 133 kg^{-1} dried.

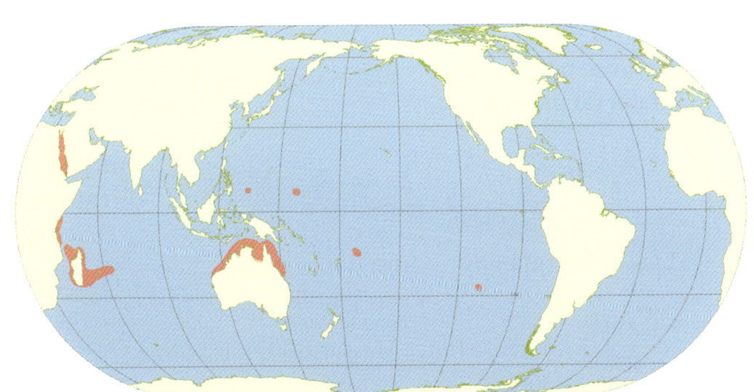

GEOGRAPHICAL DISTRIBUTION:

The species described here may have a rather patchy, but wide distribution through the Indo-Pacific. It probably occurs in more localities than presented here. It is known from Réunion, probably also in the Comoros, Madagascar and other places in the western Indian Ocean.

Stichopus naso Semper, 1868

COMMON NAMES: Because *Stichopus naso* continues to be confused with *S. horrens*, the species has been traded under the same common name of the latter species in areas where it occurs.

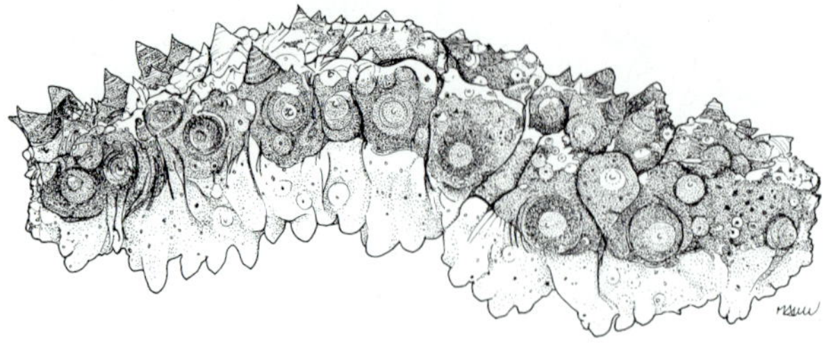

DIAGNOSTIC FEATURES: Dorsally, yellowish-tan and mottled with brown, or uniformly light brown. Laterally, somewhat lighter. Ventral surface with a brown central longitudinal band running between the rows of podia. Tips of podia and dorsal papillae are dark brown. Small specimens nearly uniformly grey, sometimes with a pair of reddish dorsolateral papillae. Trapezoidal to rectangular in cross-section; fission products appear truncate anteriorly or posteriorly. Dorsal surface lightly arched with squat, conical dorsolateral papillae. Individuals of this species are usually relatively small. Numerous, large podia arranged in longitudinal rows occur on the ventral surface. The mouth is ventral with 18–20 tentacles. Anus terminal, unguarded by papillae.

Ossicles: Tentacles with spiny rods that can bifurcate distally, 150–620 µm long, as well as C-shaped rods, 25–65 µm long. Dorsal body wall with tables, rosettes and C-shaped rods: tables with disc approximately 25 µm across, perforated by 4 central and 4–8 peripheral one; spire ending in a crown of spines resembling a Maltese cross; C-shaped rods 60–180 µm long. Ventral body wall with similar ossicle assemblage but C-shaped rods are smaller, 60–110 µm long and rosettes, 20–25 µm long, are more abundant. Ventral and dorsal podia with tables similar to those of the body wall, but also with larger ones with discs perforated with up to 20 holes, narrow and spiny rods, 200–400 µm long, rosettes, and perforated plates 100–160 µm long with spiny edges.

Processed appearance: Probably similar in size and appearance to *S. horrens*. Prominent lateral papillae on the ventral margins should be evident in processed specimens.

Remarks: Often mistaken for *S. horrens*, *S. monotuberculatus* or *S. quadrifasciatus*. The genetic and morphological boundaries of these four *Stichopus* are currently uncertain and likely to change in future.

Size: Fresh length from 10 to 20 cm, hence, probably from 100 g to 200 g fresh weight.

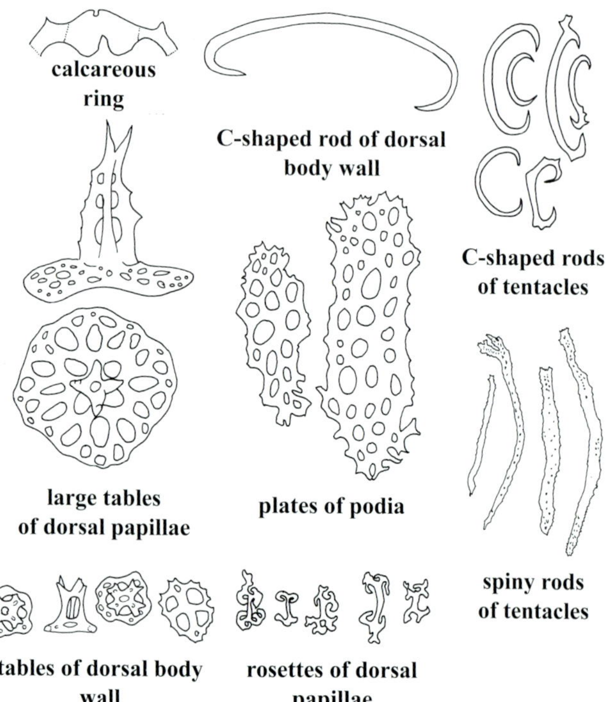

(after Massin, 2007)

HABITAT AND BIOLOGY: In shallow water from 1 to 20 m, usually on finer sediments, often associated with seagrass where it lies exposed both day and night, unlike the similar *S. horrens*, which hides under rubble during the day. Exhibits a characteristic undulating motion in an apparent effort to escape when prodded. This species, but not *S. horrens* reproduces asexually via fission, at least in the Great Barrier Reef, Australia.

LIVE (photo by: S.W. Purcell)

EXPLOITATION:

Fisheries: Exploitation is moderate in Madagascar. As this species is seldom recognized in the bêche-de-mer trade, the full extent of exploitation is uncertain.

Regulations: Because of its similarity in appearance to *S. horrens*, this species may be inadvertently regulated along with the latter species.

Human consumption: Mostly, the reconstituted body wall (bêche-de-mer) is consumed by Asians. Its intestine and/or gonads are eaten in traditional diets; local consumption is the same as for *S. horrens*.

Main market and value: It is apparently sold along with *S. horrens* in Hong Kong China SAR, China and the Philippines, where the latter species is sold at USD39 kg^{-1} dried.

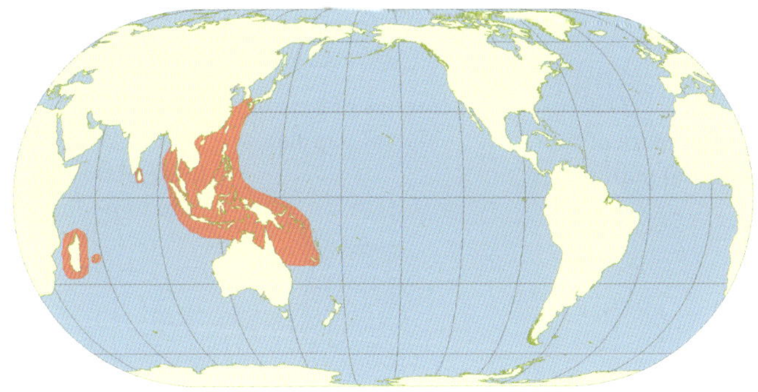

PROCESSED (photo by: C. Dissanayake)

GEOGRAPHICAL DISTRIBUTION: Wide Indo-Pacific distribution. It is recorded from the Philippines to Madagascar, including Mauritius, Sri Lanka, Thailand, Indonesia, Borneo, Australia (from the northwestern through southeastern coasts), New Caledonia, south China Sea, China, Japan, and Papua New Guinea.

COMMON NAMES: Curryfish, Ocellated sea cucumber or Hanginan (Philippines), Ñoät ngaän (Viet Nam).

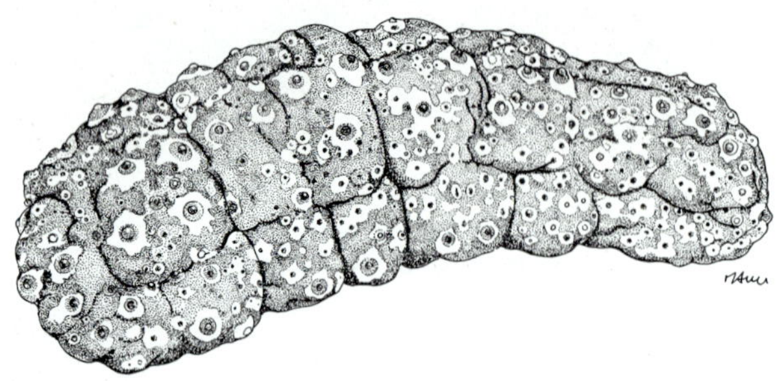

DIAGNOSTIC FEATURES: Dorsal surface is yellow or yellowish-orange with prominent, large, circular, greenish-grey, wart-like papillae that are white around the base. The large papillae occur in four rows and are arranged in a zig-zag pattern. Dorsal surface rounded. Ventral surface is flattened and whitish yellow. Podia on the ventral surface are numerous and greenish-brown, only on the ambulacral areas, and terminating with large suckers (up to 1.5 mm diameter). Mouth is ventral. Anus is terminal with no teeth.

Ossicles: Tentacles with spiny rods, 40–600 µm long. Dorsal body wall with tables, 25–40 µm across, rosettes, 20–40 µm long and C-Shaped rods, 155–175 µm long. Ventral body wall with similar ossicles, but with smaller C-shaped rods, 40–75 µm long. Ventral podia with large perforated plates, 140–265 µm long, C-shaped rods, 55–65 µm long, reduced tables, 25–50 µm across, and rods, 230–500 µm long, most of them with large central perforated process. At the base of the dorsal papillae, rosettes and C-shaped rods; at the top C-shaped rods, tables, rosettes, small rods, perforated plates and curved rods with central perforated process.

Processed appearance: Information lacking.

Remarks: In comparison with *Stichopus herrmanni*, Massin *et al.* (2002) noted that *S. ocellatus* has smaller C-shaped rod ossicles of the ventral body wall and larger (60–100 µm) perforated plates at the top of the dorsal papillae.

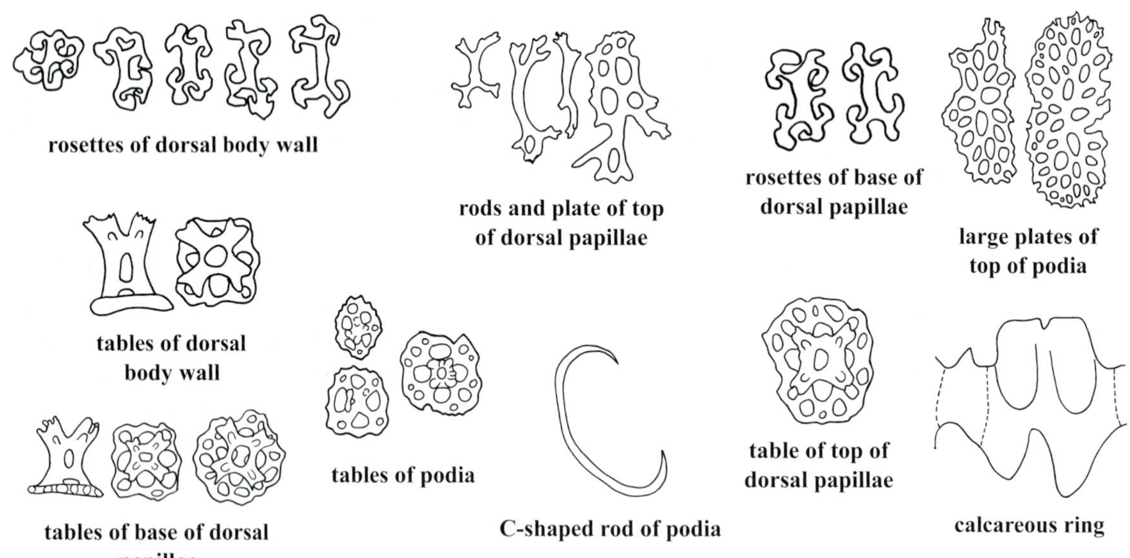

rosettes of dorsal body wall

rods and plate of top of dorsal papillae

rosettes of base of dorsal papillae

large plates of top of podia

tables of dorsal body wall

tables of podia

C-shaped rod of podia

table of top of dorsal papillae

calcareous ring

tables of base of dorsal papillae

(after Massin *et al.*, 2002)

Size: Maximum size at least 33 cm. Average fresh weight from 179 g (Viet Nam) to 1 310 g (Australia); average fresh length from 23 cm (Viet Nam) to 29 cm (Australia).

HABITAT AND BIOLOGY: *S. ocellatus* occurs mostly in seagrass beds on sandy or muddy-sand substrata on nearshore reef flats and sandflats, bivalve beds and mixed seagrass-algae beds. Massin *et al.* (2002) note that this species is often found associated with *S. herrmanni* in Malaysia and Papua New Guinea, so it may have some similar habitat preferences to that species. Its reproductive biology is unknown.

LIVE (photo by: S.M. Volkenhauer)

PROCESSED (photo by: L.B. Concepcion)

EXPLOITATION:
Fisheries: This species is primarily fished artisanally (e.g. Philippines, Malaysia, Viet Nam), but some collection at semi-industrial scales may occur in Australia. It is harvested by hookah divers (south Viet Nam, Australia) and by free diving and hand collecting (e.g. Philippines). In Australian waters, it is quite uncommon and collected only in minor quantities in Torres Strait and the Great Barrier Reef. Where it is exploited, this species is collected alongside other closely related species (*S. vastus* and *S. herrmanni*).
Regulations: In Australia, it is managed by means of permits (limited entry), no-take reserves, a combined global quota, and it is part of a rotational harvest strategy. In countries of Southeast Asia where it is fished, there are often no regulations on exploitation of this species.
Human consumption: Mostly, the reconstituted body wall (bêche-de-mer) is consumed by Asians.
Main market and value: Hong Kong China SAR. It has been traded recently at USD35–58 kg^{-1} dried in the Philippines. Wholesale prices in Guangzhou were up to USD111 kg^{-1} dried.

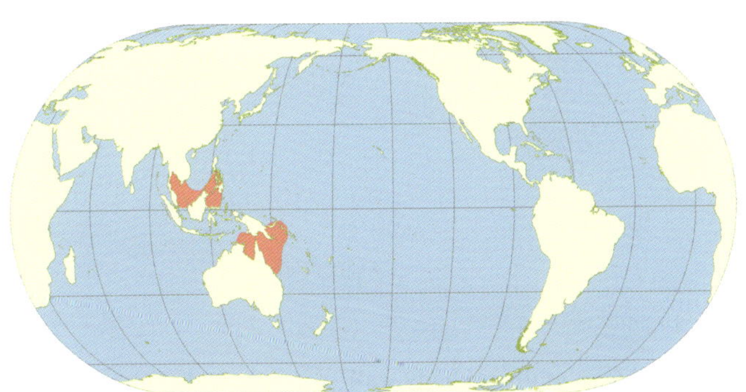

GEOGRAPHICAL DISTRIBUTION:
Papua New Guinea, northern Great Barrier Reef and Torres Strait (Australia), Malaysia, the Philippines, Viet Nam, probably Indonesia.

COMMON NAMES: Unknown.

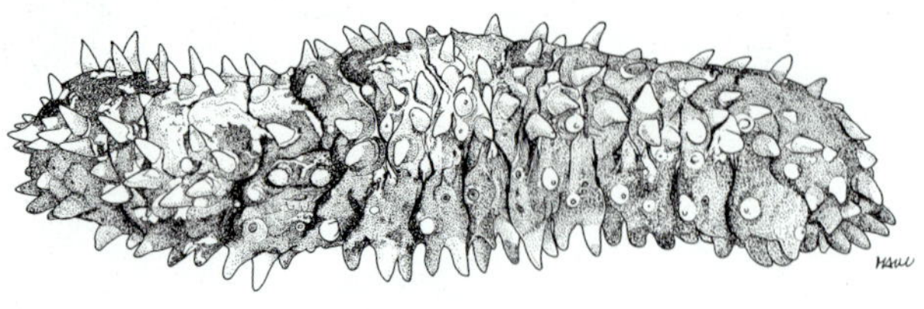

DIAGNOSTIC FEATURES:

Stichopus pseudohorrens is a large species. Coloration is brownish-yellow to rosy red with darker mottling. The body is highly arched dorsally and flattened ventrally, making it squarish to trapezoidal in cross-section. Very long, slender, conical papillae are present dorsally, especially on the upper surface of the body; the dorsal papillae are generally darker in colour than the body wall. Long papillae also occur on the lateral margins of the ventral surface. The mouth is ventral, with 20 long dark brown tentacles, and surrounded by large papillae. The anus is terminal.

Ossicles: Tentacles with spiny rods of various sizes, up to 875 µm long. Dorsal body wall with tables and C-shaped rods and rosettes. Tables comprise two types: (i) relatively few small ones, discs 50–90 µm across, perforated by 4 central and many peripheral holes, and (ii) very large tack-like ones, roughly 100 µm across, with the spire with spines. Ventral body wall with tables, C-shaped rods and rosettes: tables of one type only, disc 40–60 µm across, rim undulating, perforated by 4 central holes and few peripheral ones; spire ending in a spiny crown. Ventral podia and papillae with tables and huge rods, with a medial enlargement that can be perforated.

Processed appearance: Not available.

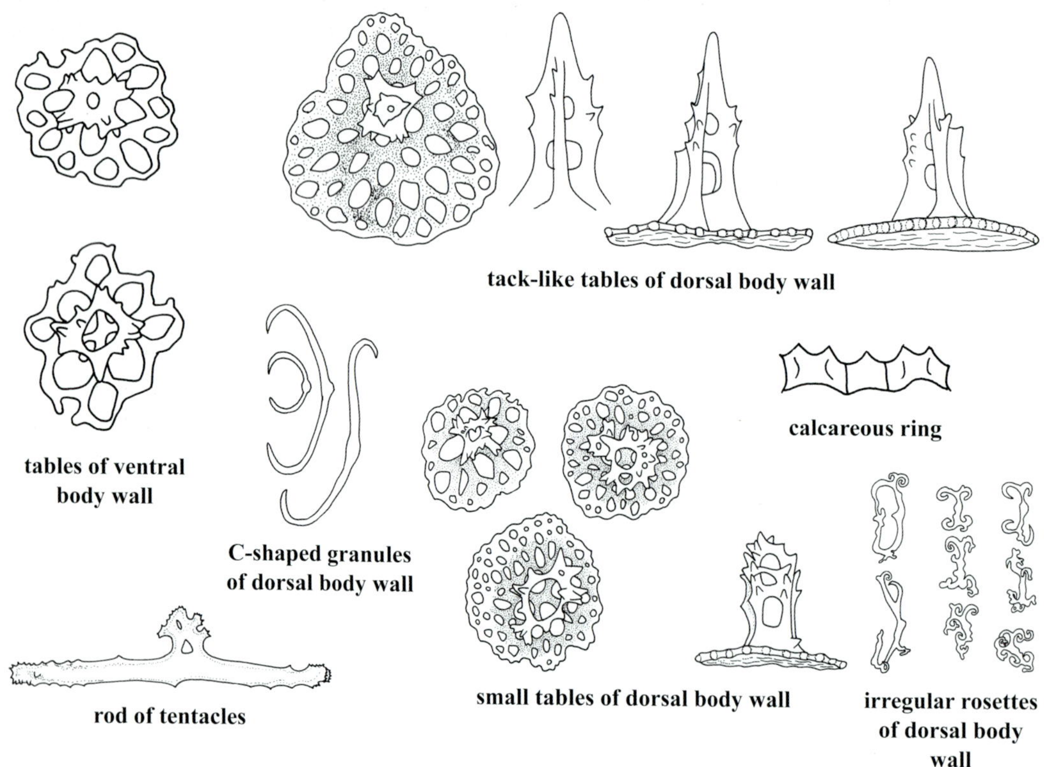

(after Cherbonnier, 1967)

Remarks: Although this species has been recorded from New Caledonia by Féral and Cherbonnier (1986), those individuals differ from Cherbonnier's (1967) original description in several ways. Therefore, populations in the Pacific may represent a different, undescribed species. For this reason, Purcell, Gossuin and Agudo (2009) referred to one individual found in New Caledonia as *Stichopus* sp. type *pseudohorrens*. The occurrence in the south China Sea needs to be validated with a voucher specimen and taxonomic description.

Size: Probably attains 50 cm and more than 3 or 4 kg in weight.

LIVE (photo by: D. VandenSpiegel)

HABITAT AND BIOLOGY: In the Comoros, it inhabits over coral sand up to 20 m depth. Its reproductive biology is unknown.

EXPLOITATION:

Fisheries: Populations in Southeast Asia and the Pacific may represent a different, as yet undescribed, species. It is probably fished artisanally and semi-industrially in localities in the western Indian Ocean.

Regulations: None.

Human consumption: Mostly, the reconstituted body wall (bêche-de-mer) is consumed by Asians. It may be included in shipments of other species, such as *Thelenota ananas*.

Main market and value: Not available.

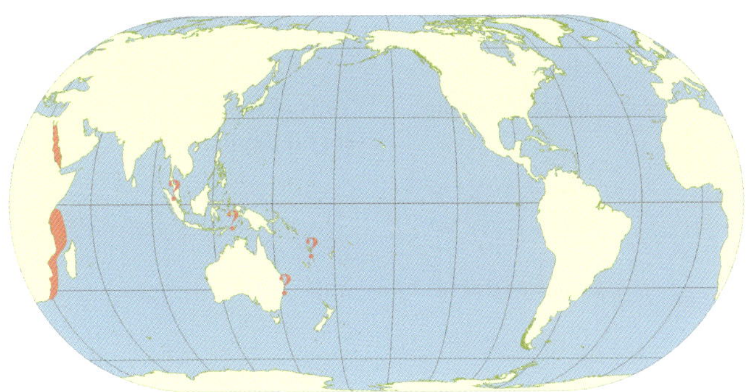

GEOGRAPHICAL DISTRIBUTION: The Comoros (off East Africa), Tanzania, South Africa and the Red Sea. Reports from the western central Pacific probably represent a separate species, and further accounts should be validated taxonomically.

Stichopus vastus Sluiter, 1887

COMMON NAMES: Curryfish (Australia), Zebrafish (India), Mul attai (India), Hanginan (Philippines).

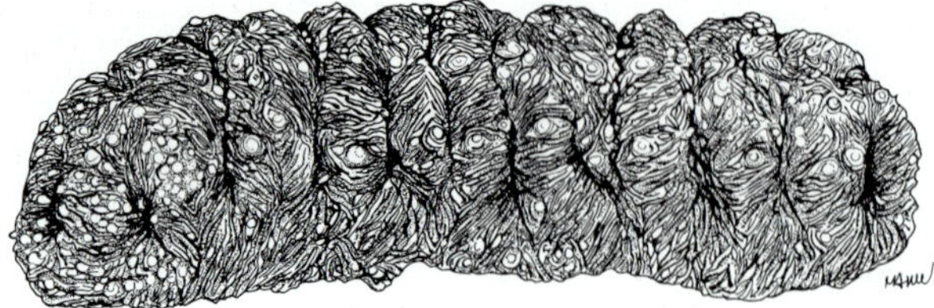

DIAGNOSTIC FEATURES: Coloration is variable from goldish-yellow, to brownish-yellow, or reddish, olive green or greyish-green. Fine, dark, discontinuous lines surround the base of large papillae on the dorsal surface. The large, wart-like papillae are present in 5–6 rows on the upper dorsal surface and along the lateral margins of the ventral surface. Smaller wart-like papillae occur all over the dorsal surface. Deep transverse wrinkles may be present dorsally. Ventral surface is brown and the interambulacral areas are yellow-orange. The body is highly arched dorsally and flattened ventrally, and may be squarish in cross-section. Large podia are numerous along the ambulacra of the ventral surface. The mouth is ventral with 18–20 tentacles, surrounded by a collar of papillae. The anus is terminal, without teeth.

Ossicles: Tentacles with curved rods with spiny extremities, 60–695 µm long. Dorsal and ventral body wall with tables, 25–40 µm across, rosettes, 15–40 µm long, and C-shaped rods, 40–95 µm long. Tables have round to quadrangular, smooth discs, perforated by 4 central and 4–10 peripheral holes; spire ends in a wide Maltese crown. Ventral podia with tables, 30–55 µm across, with reduced or no pillars and rods, 250–450 µm long, that can have a very large central perforated process. Top of dorsal papillae with large tables, 30–55 µm across, with quadrangular to ovoid disc, smooth, perforated by 4 central and 4–25 peripheral holes; spire ending in a narrow (½ to ⅓ of disc diameter) crown of spines.

Processed appearance: Light brown in colour and cylindrical in shape. Dorsal surface textured with short papillae evident. Dark lines should be visible on the dorsal surface.

Remarks: The body wall of the animals may disintegrate when handled and held out of water for a long time.

Size: Average fresh weight from 1 000 g (India) to 1 700 g (Australia); average fresh length from 33 cm (Australia) to 35 cm (India).

HABITAT AND BIOLOGY: This species is found on inshore reefs edges on sand, coral rubble or muddy sand in shallow waters, generally to about 8 m depth. On the Great Barrier Reef (Australia), it may be found on sandy or coral rubble substrates at the base of semi-sheltered reefs. Its reproductive biology is unknown.

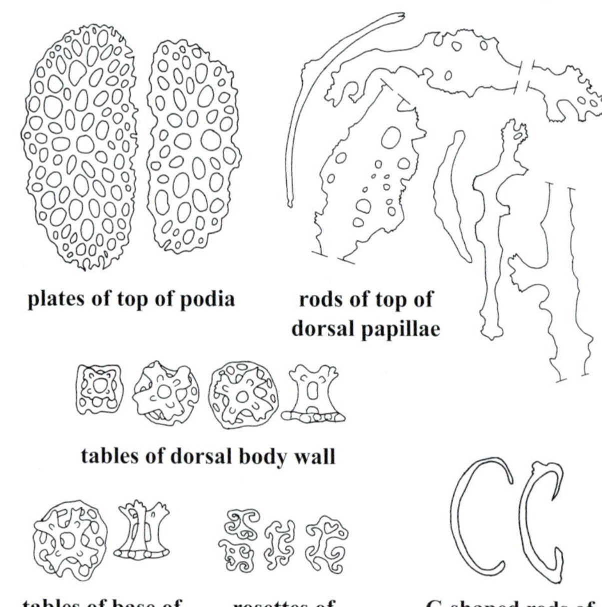

(after Massin *et al.*, 2002)

EXPLOITATION:

Fisheries: This species is exploited in artisanal (Micronesia [Federated States of]) and semi-industrial fisheries (Australia). On the Great Barrier Reef (Australia), it is less common than *Stichopus herrmanni* and is occasionally collected by divers using hookah. In the western central Pacific, it is commercially harvested in Palau, Micronesia (Federated States of) and Torres Strait (Australia). This species was harvested in Papua New Guinea, Solomon Islands and parts of Vanuatu prior to national moratoria. The subsistence fishery in Palau targets the gonads and/or intestines. It is believed to be heavily exploited in Indonesia. This species is probably collected and processed alongside *S. herrmanni* in places where it is fished.

Regulations: In Australia, exploitation of this species is regulated by fishing permits (limited entry), no-take reserves, a combined TAC, and is subject to a rotational harvest strategy.

Human consumption: Mostly, the reconstituted body wall (bêche-de-mer) is consumed by Asians.

Main market and value: Hong Kong China SAR. It has been traded recently at USD35–58 kg^{-1} dried in the Philippines.

LIVE (photo by: S.W. Purcell)

PROCESSED (photo by: S.W. Purcell)

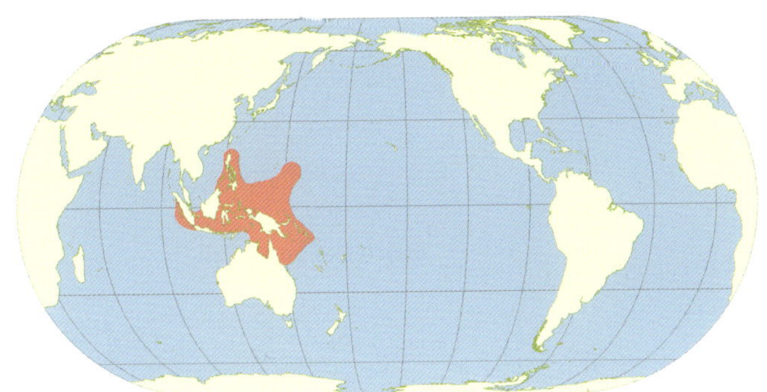

GEOGRAPHICAL DISTRIBUTION:

Indonesia, the Philippines, Papua New Guinea, Palau Islands, Yap (Federated States of Micronesia) and northeastern Australia. Although also reported from Uri, Vanuatu, it does not appear to occur in New Caledonia.

Thelenota ananas (Jaeger, 1833)

COMMON NAMES: Prickly redfish (**FAO**), Holothurie ananas (**FAO**), Hải sâm lựu, Đồn đột lựu, Ñoät ñieàu (Viet Nam), Zanga borosy, Rasta (Madagascar), Barbara (Mauritius), Ananas attai (India), Subinho (Mozambique), Tinikan, Talipan, Taripan, Pinya-pinya (Philippines), Abu mud (Eritrea), Pandan (Indonesia), Spinyo mama (Tanzania), Baba (Zanzibar, Tanzania), Teburere (Kiribati), Pulukalia (Tonga), Sucudrau (Fiji).

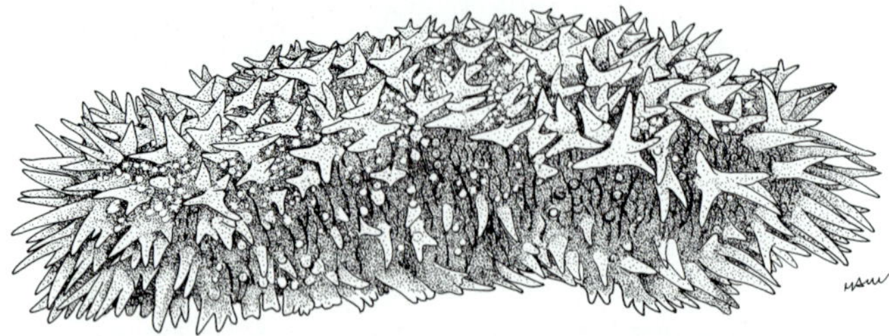

DIAGNOSTIC FEATURES: Colour variable dorsally from reddish-orange to brown or burgundy. Dorsal surface is covered in very large papillae, which may be long and conical or star-shaped on a short stalk or somewhat branched. Ventral surface is light pink to red, with brown to pink podia more abundant on the radii. Body is firm and rigid; arched dorsally and flattened ventrally. Body wall is thick. Mouth ventral with 20 large, brown tentacles, surrounded by conical papillae. Anus is terminal, and often hidden by large papillae. Cuvierian tubules are absent.

Ossicles: Tentacles with large plates, 135 µm long and 95 µm wide, as well as some smaller rods. Dorsal body wall with dichotomously branched rods, which are slightly spiny, 40–80 µm long and countless miliary granules, 1.5–4 µm long. Ventral body wall with similar, though smooth, rods. Ventral podia with large plates, 75–135 µm long, and rods similar to those of the body wall. Dorsal papillae with branched rods similar to those of the body wall as well as slightly curved, spiny rods, up to 155 µm long.

Processed appearance: Relatively elongate and brown to black in colour. The dorsal surface is covered with brown to black-brown spikes, often in a star shape. The ventral surface is granular and lighter brown. One long cut along the ventral surface. Common processed size 20–25 cm.

Size: Maximum length 80 cm; average length generally about 45 cm. Average fresh weight: 1 000 g (Mauritius), 2 000 g (Réunion), 2 500 g (Papua New Guinea), 2 600 g (New Caledonia), 3 000 g (India); average fresh length: 34 cm (New Caledonia), 35 cm (Mauritius), 45 cm (India and Papua New Guinea), 45–80 cm (Réunion).

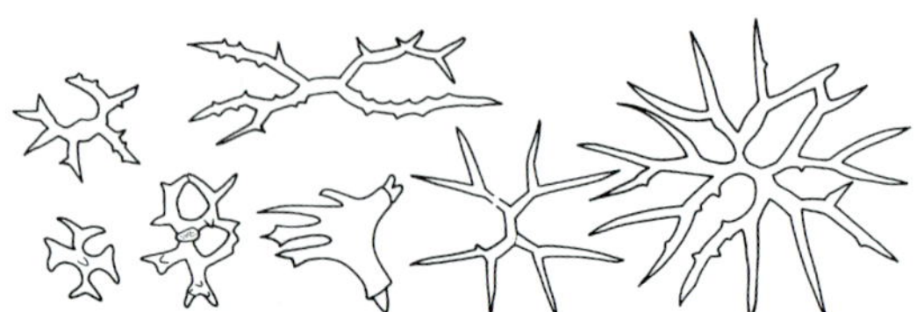

ossicles of body wall

HABITAT AND BIOLOGY: In the western central Pacific, it prefers reef slopes and passes, hard bottoms with large coral rubble and coral patches in waters between 1 and 25 m. In the Africa and Indian Ocean region, this species prefers coral slopes over hard substratum between 5 and 35 m. It attains size-at-maturity at 1 200 g and reproduces annually during the warm season. In Guam, it reproduces almost all year long, with the exception of March, September and October; and in New Caledonia, it has an annual reproductive cycle from January to March. It has a late sexual maturity.

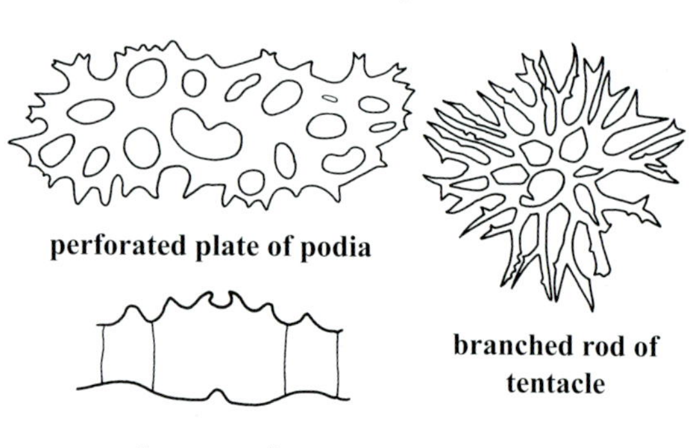

perforated plate of podia

calcareous ring

branched rod of tentacle

(after Féral and Cherbonnier, 1986)

EXPLOITATION:

Fisheries: Exploited in artisanal (e.g. Mozambique, Tonga), semi-industrial (e.g. Viet Nam, New Caledonia) and industrial fisheries (e.g. Australia). It is exploited in all fisheries throughout its Indo-Pacific distribution, and is one of the most valuable species. In Viet Nam and Australia, it is harvested by hand by divers using hookah. It was heavily exploited in Papua New Guinea, Solomon Islands and Vanuatu prior to national moratoria. It is harvested in subsistence fisheries in Samoa and Cook Islands. In Asia, *T. ananas* is exploited in China, Japan, Malaysia, Thailand, Indonesia (heavily fished), the Philippines and Viet Nam. It is fished in Madagascar, Kenya, Maldives, Eritrea and Seychelles. Juveniles are also sometimes collected for sale in the aquarium trade.

LIVE (photo by: S.W. Purcell)

PROCESSED (photo by: J. Akamine)

Regulations: In Papua New Guinea, there was a minimum size limit of 25 cm live and 10 cm dry. In New Caledonia, the legal minimum length is 45 cm live and 20 cm dry, collection using compressed air diving is prohibited and there are no-take reserves. The minimum legal length in Torres Strait, Northern Territory and Western Australia is 30 cm live, whereas it is 50 cm live on the Great Barrier Reef. On the Great Barrier Reef, there is a TAC, limited entry and permits, and this species is subject to a rotational harvest strategy.

Human consumption: The reconstituted body wall (bêche-de-mer) is consumed by Asians. In some localities in the Pacific, it is consumed either whole or its intestine and/or gonads as delicacies or as protein in traditional diets or in times of hardship (i.e. following cyclones).

Main market and value: China, Singapore, Hong Kong China SAR, Ho Chi Minh City (Viet nam) for further export to Chinese markets. It is exported at about USD50 kg^{-1} dried in Viet Nam. It has been traded recently at USD35–63 kg^{-1} dried in the Philippines. It was previously sold at USD12–17 kg^{-1} (dried) in Papua New Guinea. In New Caledonia, it is exported for USD40–50 kg^{-1} dried. In Fiji, fishers receive USD11–18 per piece fresh. Prices in Guangzhou wholesale markets ranged from USD22 to 184 kg^{-1} dried.

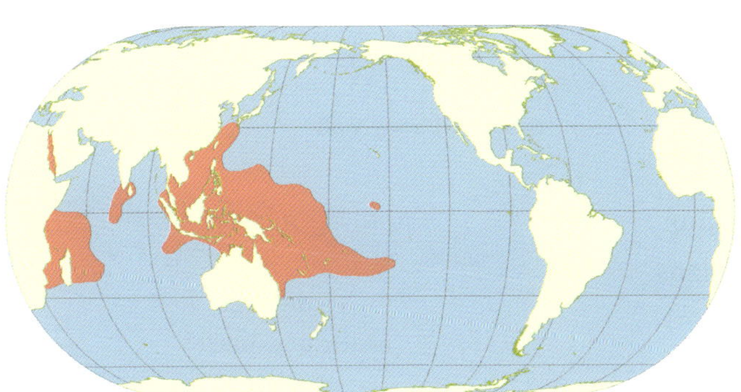

GEOGRAPHICAL DISTRIBUTION:

Red Sea, Mascarene Islands, Maldives, East Indies, North Australia, the Philippines, Indonesia, China and southern Japan, and islands of the Central Western Pacific as far east as French Polynesia.

Thelenota anax Clark, 1921

COMMON NAMES: Amber fish (**FAO**), Legs (Philippines), Le géant (New Caledonia), Saieniti (Tonga), Dri-volavola (Fiji), Teaintoa (Kiribati).

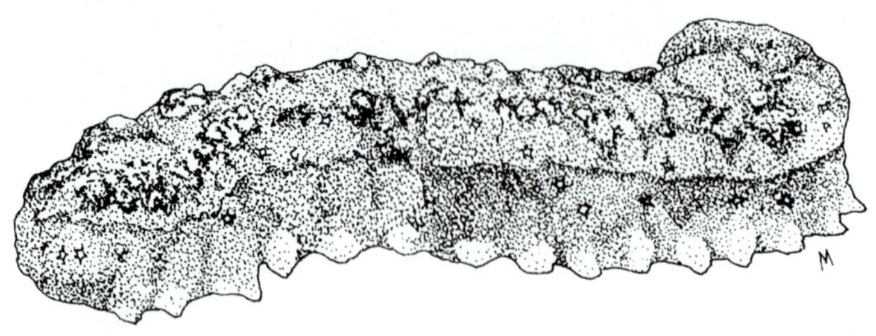

DIAGNOSTIC FEATURES:
Thelenota anax is a very large species. Colour varies from creamy white beige to grey or light brown with dark brown and/or reddish spots and blotches dorsally. Those in the Indian Ocean may lack the reddish blotches. Numerous, light coloured, wart-like bumps occur mostly in rows along either side of the dorsal surface. Large, white papillae are located along the ventro-lateral margins. It has a thick body wall. Body is rather quadrangular in cross-section. The flat ventral surface is densely covered with fine, long podia. The mouth is ventral with 18–20 peltate tentacles. The anus is terminal to subdorsal. Cuvierian tubules are absent.

Ossicles: Tentacles with nodulous and branched rods, and perforated plates, 80–100 µm long. Ventral and dorsal body wall with dichotomously branched rods, 70–100 µm long, pseudo-tables and an infinite number of miliary granules only a few µm across. Ventral podia with rods of various form; straight to arms inter-joining as well as turrets. Ossicles of dorsal papillae are long rods, which can be branched and perforated; or more plate-like deposits.

Processed appearance: Relatively elongate with a squarish cross-section. Body may be various shades of brown. Dorsal surface is rough and covered with wart-like bumps. The ventral surface is grainy. A small cut across mouth or a single cut ventrally. Common processed size is 15–20 cm.

Size: Maximum length: 89 cm; average length: 63 cm. In Papua New Guinea, average fresh weight was about 3 500, and average fresh length 35 cm. At Lizard Island, Great Barrier Reef (Australia), average length about 51 cm and average drained weight about 3 340 g. Average length of 55 cm and an average weight of 4 370 g from New Caledonia.

HABITAT AND BIOLOGY: It primarily inhabits reef slopes and outer lagoons on sandy bottoms between 10 and 30 m. It may be found less commonly in shallower waters to about 4–5 m depth, and on hard bottoms or on coral rubble. Generally found at low density, the populations are usually sparse.

Its reproductive biology is unknown.

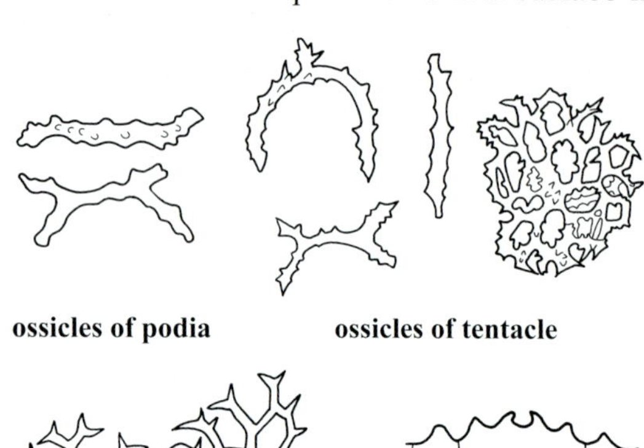

ossicles of podia   ossicles of tentacle

ossicles of body wall   calcareous ring

ossicle of papillae

(after Féral and Cherbonnier, 1986)

EXPLOITATION:

Fisheries: This species is mainly exploited artisanally. In many fisheries in the Central Western Pacific, Southeast Asia and the Indian Ocean, it is collected by hand by breath-hold divers or divers using compressed air (SCUBA or hookah). Before a national moratorium in Papua New Guinea, it was also collected by free divers using lead-bombs. In the Philippines, it is harvested by divers using hookah and SCUBA.

Regulations: In Papua New Guinea, fishery regulations included a minimun landing size of 20 cm live and 10 cm dry. On the Great Barrier Reef (Australia), there is a minimum legal length of 50 cm live, limited entry of fishers, who are licensed, a combined TAC (with other species), no-take marine reserves, and this species is subject to a rotational harvest strategy.

Human consumption: Mostly, the reconstituted body wall (bêche-de-mer) is consumed by Asians.

Main market and value: Hong Kong China SAR and other import hubs of Southeast Asia. It is sold at about US$3–4 kg^{-1} dried in some localities, but probably at much higher prices in other localities. It has been traded recently at about USD13 kg^{-1} dried in the Philippines. Prices in Guangzhou wholesale markets ranged from USD14 to 32 kg^{-1} dried.

LIVE (photo by: S.W. Purcell)

PROCESSED (photo by: S.W. Purcell)

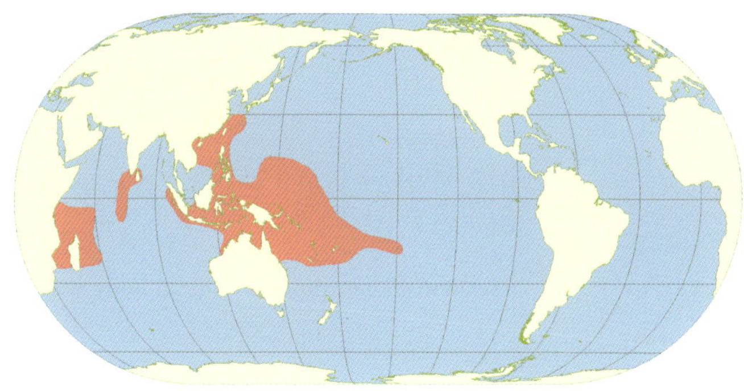

GEOGRAPHICAL DISTRIBUTION:

Tropical Indo-west Pacific. In the tropical Indian Ocean, this species is known from East Africa, the Comoros and Glorieuses Islands. It is present in much of Southeast Asia, including Indonesia, the Philippines and the south China Sea. In the tropical Pacific, from northwestern Australia to Enewetok, Guam, and the Ryukyu Islands southwards to most of the islands of the Central Western Pacific and as far east as French Polynesia.

Thelenota rubralineata Massin and Lane, 1991

COMMON NAMES: Lemonfish (Solomon Islands), Candycanefish or Candy cane sea cucumber (Philippines), red-lined sea cucumber.

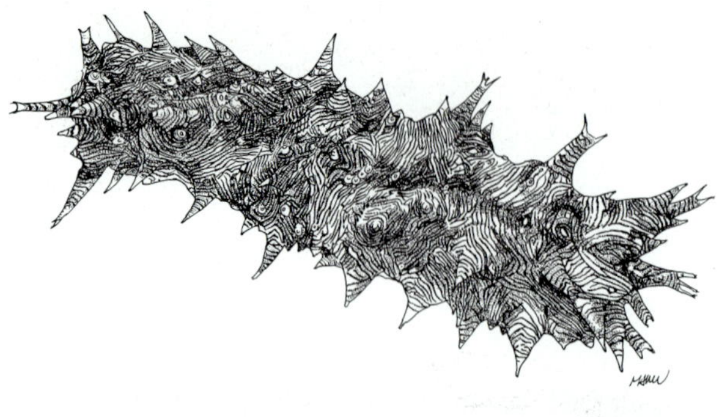

DIAGNOSTIC FEATURES: This species is whitish coloured with a striking and complex pattern of crimson lines, which form a maze-like arrangement. The crimson lines are less abundant and more irregular ventrally. The dorsal surface has two rows of 13–15 large, conical, fleshy protuberances with pointed papillae at the ends, with yellowish brown tips. The body is roughly square or trapezoid in transverse section. The posterior part of the body tapers slightly. The ventral surface is flattened and has numerous greenish-yellow or brownish-yellow podia scattered randomly. The mouth is ventral with 20 dull-red tentacles. The anus is terminal.

Ossicles: Tentacles with only rods, which are spiny or smooth, straight or curved, 10–150 μm long. Dorsal body wall with numerous miliary granules, slender dichotomously branched rods, 90–135 μm long, that are spiny, with primary, secondary, tertiary and sometimes quaternary branches, and pseudo-tables, 20–25 μm across with 4–5 short feet which are prolonged by 4–5 long spines. Ventral body wall with dichotomous rods, pseudo-tables and serpent-like granules. Ventral podia with rods, a few pseudo-tables and anastomosing plates forming the end-plate. Dorsal papillae with mostly serpent-like granules, 5–20 μm long, and a few dichotomous rods.

Processed appearance: Poorly known. From few samples: coloration is brown, body is relatively elongate with the characteristic large, pointed protuberances retained on the dorsal surface.

Size: Maximum weight about 3 kg. Average length is 30–50 cm.

HABITAT AND BIOLOGY: Found on reef slopes and spur zones.
Its reproductive biology is unknown.

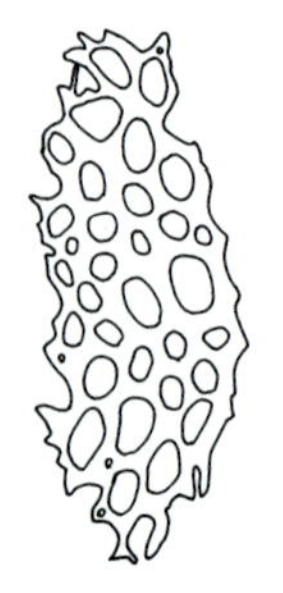

plate of end
of podia

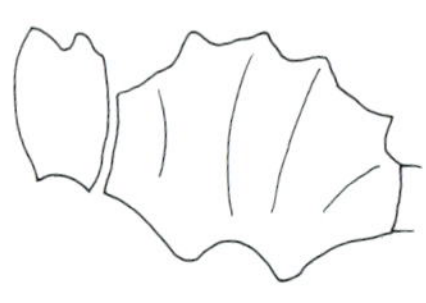

calcareous ring

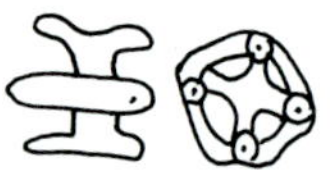

pseudo-tables of
dorsal body wall

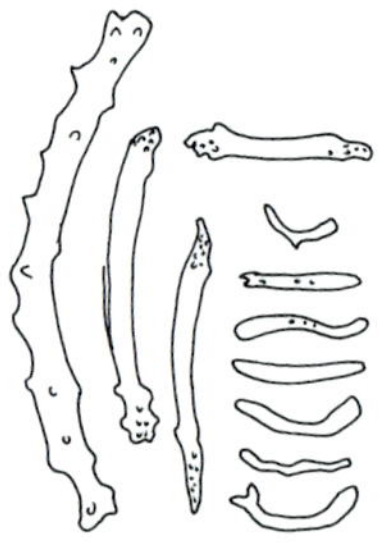

rods of tentacles

serpent-like granules of dorsal body wall

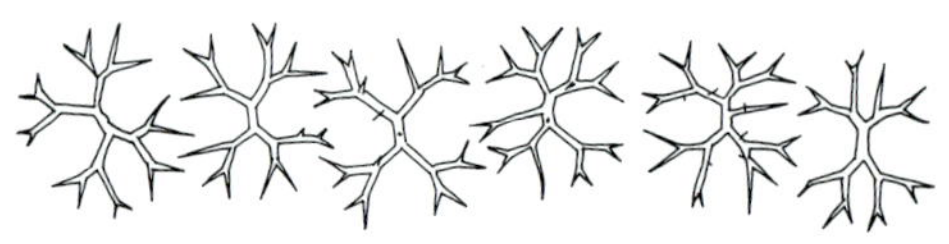

dichotomous rods of dorsal body wall

(after Massin and Lane, 1991)

EXPLOITATION:

Fisheries: In the western central Pacific it was commercially harvested in Papua New Guinea and Solomon Islands prior to moratoria.

Regulations: It seems there are few regulations for the exploitation of this species within its distributional range.

Human consumption: In the Philippines, it is a consumed by Muslims during the Ramadan season.

Main market and value: *Thelenota rubralineata* has a moderately low value by weight. It has been traded recently at about USD13 kg^{-1} dried in the Philippines.

LIVE (photo by: K. Friedman)

PROCESSED (photo by: L.B. Concepcion)

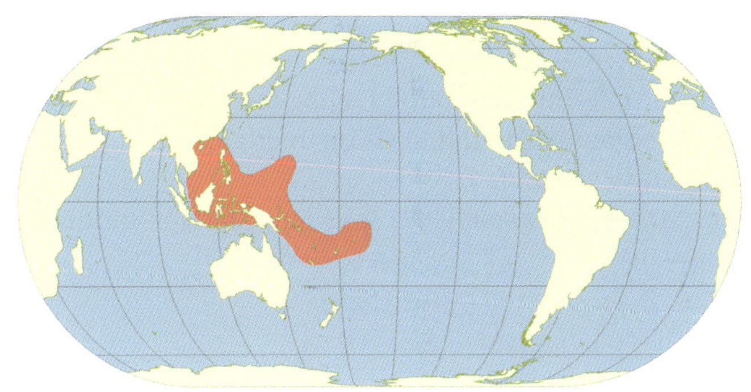

GEOGRAPHICAL DISTRIBUTION:

This species has been found in the "Coral Triangle" and extends into the Pacific Ocean. In Southeast Asia it has been recorded from Indonesia, the Philippines, east Malaysia, islands of the south China Sea; and from the Pacific region, Guam, Solomon Islands, Papua New Guinea, and it has been sighted (but rarely) in New Caledonia, and possibly Fiji.

Athyonidium chilensis (Semper, 1868)

Common names: Sea cucumber (USA), Pepino de mar, Pepino arenero, Ancoco (Mexico and Peru), Meón, Ancoco blanco (Chile).

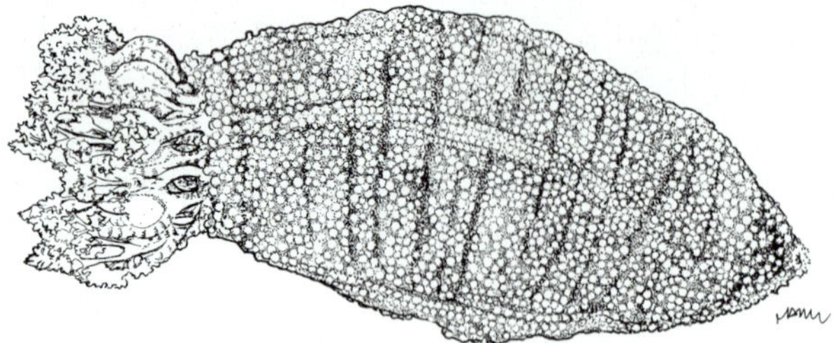

DIAGNOSTIC FEATURES: Body is brown to greyish-brown or light grey. Juveniles can be greenish. The body of this species is cylindrical and tapers gently at both ends. Podia are found in longitudinal rows along the body wall. The mouth is terminal with five pairs of greenish-black, branched dendritic tentacles arranged in two circles: five large external pairs, and five small inner pairs.

Ossicles: Tentacles of juveniles with rods. Dorsal and ventral body wall with few spiny, perforated rods that are somewhat enlarged at the extremities. Ventral podia only have an end-plate.

Processed appearance: Dark brown to near black in colour. Processed body shape is similar to the description of the live animal: cylindrical and tapers gently at both ends. Body surface somewhat rough. Common dried size 7–10 cm.

Size: Average fresh weight: 200–250 g; average fresh length: 25–30 cm.

HABITAT AND BIOLOGY: This species lives partially, or fully, buried in the intertidal zone to 7 m depth, or in areas with rocks with great quantity of organic matter. It feeds on microalgae and detritus. It is possible to find some invertebrates in its digestive tract.

It has a continuous reproductive period. In Peru, *Athyonidium chilensis* spawning starts in spring and lasts for four to six months.

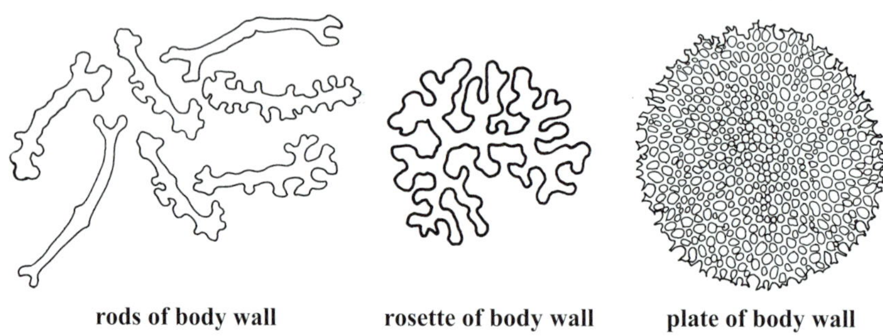

rods of body wall   rosette of body wall   plate of body wall

(source: photo L. Amaro-Rojas)

EXPLOITATION:
Fisheries: This species is harvested artisanally by hookah diving and hand collecting. It is commercially exploited in Peru and Chile. Historical information states that it was traditionally eaten in the Department of Lambayeque (Peru).
Regulations: None.
Human consumption: Mostly, the reconstituted body wall (bêche-de-mer) is consumed by Asians.
Main market and value: The United States of America, China, Mexico, Taiwan Province of China. It is sold at USD10 kg^{-1} dried.

LIVE (photo by: L. Amaro-Rojas)

PROCESSED (photo by: C. Guisado)

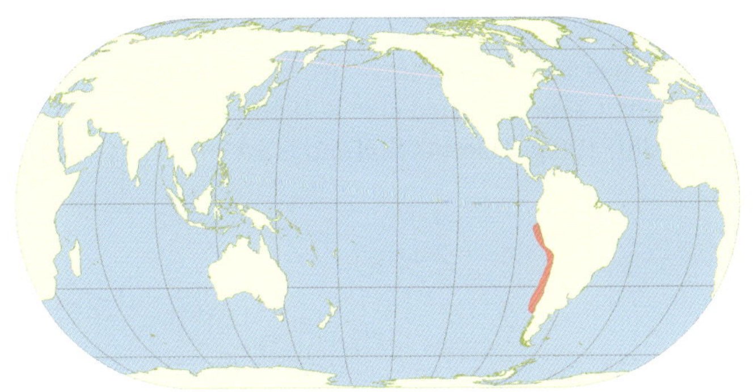

GEOGRAPHICAL DISTRIBUTION:
From Ancon (Peru) 12° 02.3'S; 75° 19.4'W) to Chiloé (Chile) (42° 48'S; 74° 21'W).

Cucumaria frondosa (Gunnerus, 1767)

COMMON NAMES: Orange-footed sea cucumber, Northern sea cucumber, Phenix sea cucumber, Pumpkin (Canada), Atlantic sea cucumber (USA and Russian Federation).

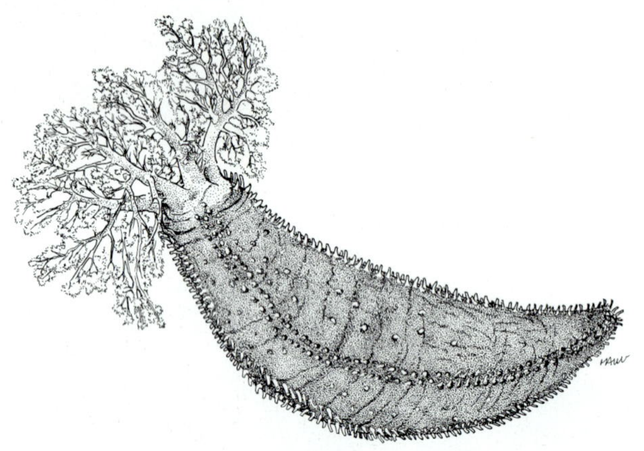

DIAGNOSTIC FEATURES: Body is mostly light or dark brown but, infrequently, individuals can be much lighter in colour, such as light orange or cream. Coloration tends to become yellowish near the mouth and tentacles. Body is cylindrical, relatively stout and barrel-shaped, slightly curved dorsally, and tapering gently at both ends. Podia are located over most of the body, primarily in five wide rows. The body contracts to become almost spherical when the animal is handled. The mouth is terminal, with 5 pairs of dendritic tentacles, which the animal extends into currents for suspension feeding. The anus is terminal.

Ossicles: Tentacles with rods or plates, 120–350 µm long. The ossicles of the body wall are perforated plates of different sizes, 200–250 µm long, triangular, quadrangular or subcircular with ragged edge; the surface of the plates is either smooth or with projections. The body wall around the anus holds larger, more spiny plates that can have a secondary spiny layer, 200–400 µm long. Ventral podia with straight, or slightly curved perforated rods, 250–300 µm long, that can be smooth or nodulous.

Processed appearance: Dark brown. The body has a slightly grainy surface with visible rows of podia, which are often more defined in the underside. Common dried size 8–9 cm.

Size: Maximum length: 50 cm. Average fresh weight: 500 g (Barents Sea and USA), 850 g (Canada); average fresh length: 25–30 cm.

HABITAT AND BIOLOGY: Coastal, offshore in the Barents Sea; coastal in Maine. It can be found in rocky or pebbly substratum and gravel bottoms from low tide mark down to about 500 m. This species often lives with its posterior end partially buried in gravel or sand. Observed in exceptional abundance on sandy bottoms, especially on the Newfoundland Grand Banks of Canada. It can also be found on rocky substrates. This species has an annual

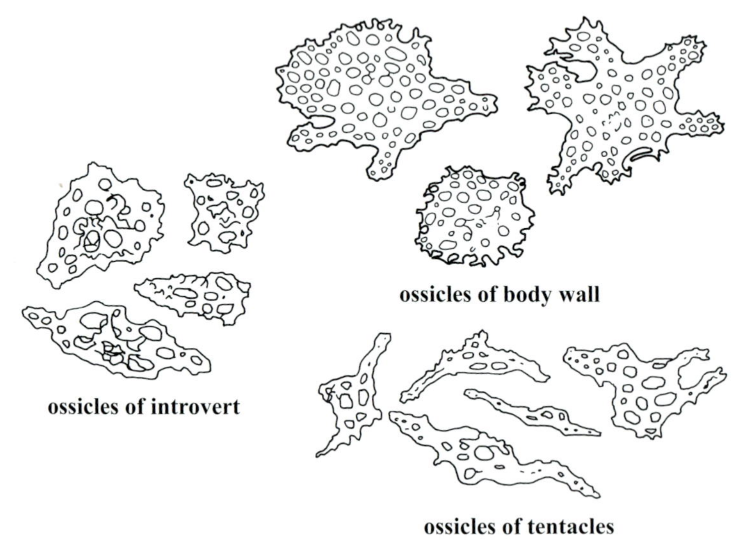

ossicles of body wall

ossicles of introvert

ossicles of tentacles

(after Levin and Gudimova, 2000)

reproductive cycle with a generally highly synchronized gamete release, between the months of February and June, depending on the region. It reaches size-at-maturity at 50 g, in the Barents Sea (Russian Federation) and at 9–12 cm in Canada.

EXPLOITATION:

Fisheries: Exploitation is industrial in Canada and semi-industrial in the Barents Sea. It is harvested by trawling (Canada) and dredging (Barents Sea and Maine).

Regulations: In Canada, there is a minimum legal landing length of 10 cm (fresh) and the exploitation is under the Emerging Fishery Policy. In the Barents Sea, there is a fishing season between August and December, while the legal fishing season in the United States of America is during autumn and winter.

Human consumption: Consumed as bêche-de-mer and as canned (undried) product. Muscle bands and the epipharyngeal 'bulb' are removed from the animals and also exported to Asian markets. It is used commercially for the preparation of traditional medicinal products

Main market and value: The market is China. In Canada, it is exported, and the fishers receive USD0.25 per animal (fresh).

LIVE (photo by: J.F. Hamel & A. Mercier)

PROCESSED (photo by: S.W. Purcell)

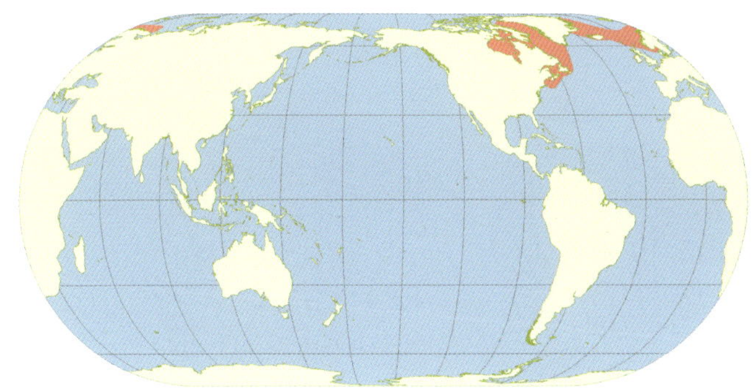

GEOGRAPHICAL DISTRIBUTION:

North Atlantic, from the Arctic to Cape Cod, and in the northern latitudes of the United Kingdom of Great Britain and Northern Ireland (Scotland and Orkney Islands). It is found in Iceland, in the Barents Sea along the coast of the Russian Federation, Scandinavia, and along the coast of Greenland.

Cucumaria japonica Semper, 1868

Common names: Japanese cucumaria, Kinko (Japan), Black sea cucumber (Canada).

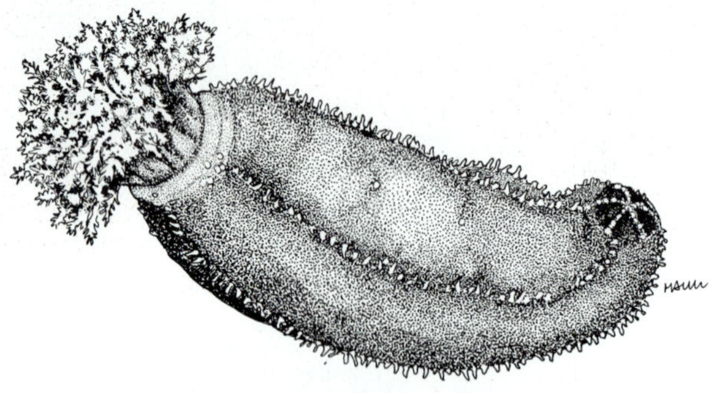

DIAGNOSTIC FEATURES: Body colour is brown to brownish-purple or greyish-purple, and in some regions the animals can be white. Podia and small papillae occur on the body in five, thin, longitudinal rows, and may be yellowish in colour. Body is cylindrical, relatively stout and barrel-shaped, curved dorsally, and tapering gently at both ends. The body contracts to become almost spherical when the animal is handled. The mouth is terminal, with 5 pairs of dendritic tentacles, which may be reddish with whitish tips. The anus is terminal.

Ossicles: Similar in size and shape to *Cucumaria frondosa*, i.e. with irregular perforated plates with spiny margins in the body wall and tentacles, some bearing knobs or short spines on their surface. However, unlike *C. frondosa* according to Semper, *C. japonica* also has large discoid ossicles positioned radially near the opening of the cloaca.

Processed appearance: Body is dark tan to brown in colour, with lighter dots occuring in five rows along the body. The mouth may be absent from dried specimens. The body is stout and tapers thin at the anus.

Remarks: Can be mistaken for *C. frondosa* and some authors consider *C. japonica* to be a subspecies of *C. frondosa*. However, Levin and Gudimova (2000) argued that these are two distinct species. Therefore, these are presented as two species in this book.

Size: Maximum weight 1.5–2 kg; maximum length: 40 cm. Average fresh weight 500 g; average fresh length: 20 cm.

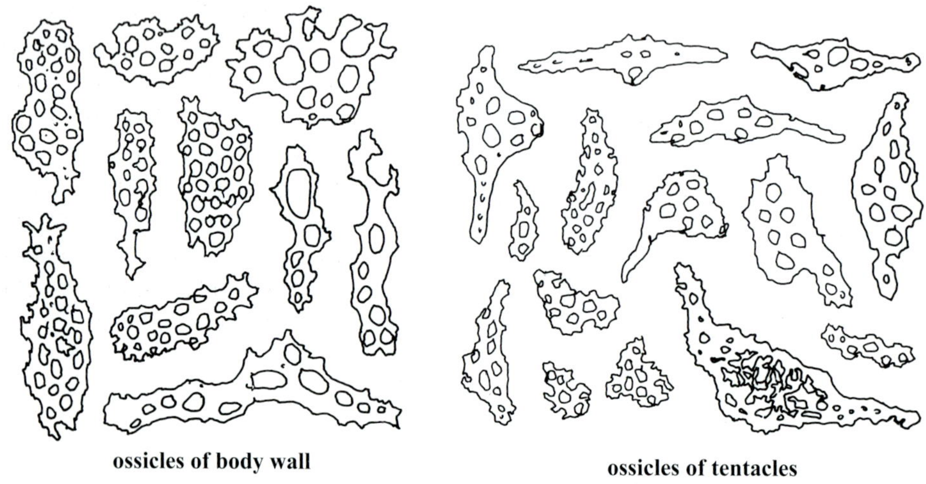

ossicles of body wall   ossicles of tentacles

(after Gudimova, 1998)

HABITAT AND BIOLOGY: This species inhabits from the intertidal zone to about 300 m depth with highest density between 30 and 60 m. Adults found on various substrates (gravel, shell debris, rock and mud), while juveniles are believed to occur more in kelp forests.

It reproduces bi-annually in April–June and September–October.

LIVE (photo by: N. Sanamyan)

EXPLOITATION:

Fisheries: This species is exploited in industrial and semi-industrial fisheries. It is commercially harvested by trawling in Japan along the coast of the Sea of Japan and north of Ibaraki Prefecture to Hokkaido. It is also fished along the Pacific coast of the Russian Federation, although the fishery information is unclear.

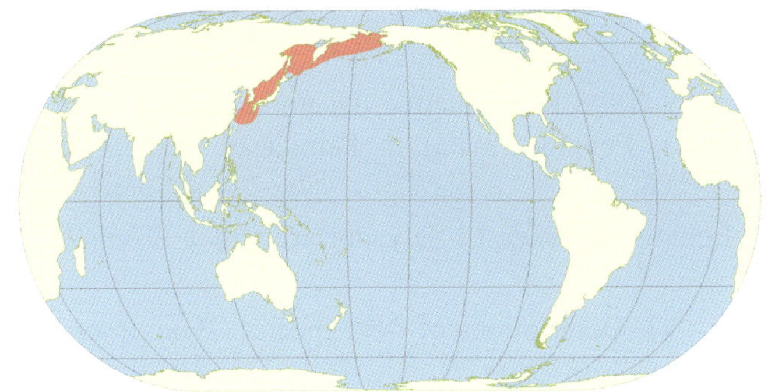

PROCESSED (photo by: J. Akamine)

Regulations: Exploitation is managed by a seasonal fishing closure, no-take marine reserves and gear limitations.

Human consumption: Mostly, the reconstituted body wall (bêche-de-mer) is consumed by Asians, or the undried body wall is sold frozen.

Main market and value: The republic of Korea and China are the main markets.

GEOGRAPHICAL DISTRIBUTION:
It inhabits the northeastern region of the Yellow Sea, the northeastern coast of the Honshu Island, the coast of the Russian Federation in the Sea of Japan, the Sea of Okhotsk, the Kuril Islands, the Kamchatka Peninsula and the Bering Sea. However, it is argued that *C. okhotensis* is indeed present in the Kamchatka and Kuril Islands areas instead of *C. japonica*.

BIBLIOGRAPHY

Abdel Razek, F.A., Abdel Rahman, S.H., Mona, M.H., El-Gamal, M.M. & Moussa, R.M. 2007. An observation on the effect of environmental conditions on induced fission of the Mediterranean sand sea cucumber, *Holothuria arenicola* (Semper, 1868) in Egypt. *SPC Beche-de-mer Inf. Bull.*, 26: 33–34.

Alfonso, I., Frías, M.P., Aleaga, L. & Alonso, C. 2004. Current status of the sea cucumber fishery in the south eastern region of Cuba. *In* A. Lovatelli, C. Conand, S. Purcell, S. Uthicke, J.-F. Hamel & A. Mercier, eds. *Advances in sea cucumber aquaculture and management*, pp. 151–159. FAO Fisheries Technical Paper No. 463. Rome, FAO. 425 pp.

Archer, J.E. 1996. Aspects of the reproductive and larval biology and ecology, of the temperate holothurian *Stichopus mollis* (Hutton). M.Sc. thesis, Univ. of Auckland, New Zealand. 189 pp.

Asha, P.S. & Muthiah, P. 2007. Reproductive biology of the commercial sea cucumber *Holothuria spinifera* (Echinodermata: Holothuroidea) from Tuticorin, Tamil Nadu, India. *Aquaculture International*, 16(3): 231–242.

Aumeeruddy, R. 2007. Sea cucumbers in Seychelles. *In* C. Conand & N. Muthiga, eds. *Status report of sea cucumber fisheries in the western Indian Ocean*. WIOMSA.

Aumeeruddy, R. & Payet, R. 2004. Management of Seychelles sea cucumber fishery: status and prospects. *In* A. Lovatelli, C. Conand, S. Purcell, S. Uthicke, J.-F. Hamel and A. Mercier, eds. *Advances in sea cucumber aquaculture and management*, pp. 239–246. FAO Fisheries Technical Paper No. 463. Rome, FAO. 425 pp.

Aumeeruddy, R. & Conand, C. 2008. Seychelles: a hotspot of sea cucumber fisheries in Africa and the Indian Ocean region. *In* V. Toral-Granda, A. Lovatelli & M. Vasconcellos, eds. *Sea cucumbers. A global review of fisheries and trade*, pp. 195–209. FAO Fisheries and Aquaculture Technical Paper No. 516. Rome, FAO. 317 pp.

Aumeeruddy, R., Skewes, T., Dorizo, J., Carocci, F., Coeur de Lion, F., Harris, A., Henriette, C. & Cedras, M. 2005. *Resource assessment and management of Seychelles sea cucumber fishery.* FAO Project Number: TCP/Seychelles/2902(A), November 2005. 49 pp.

Baine, M. & Choo, P.S. 1999. Sea cucumber fisheries and trade in Malaysia. *In* M. Baine, ed. *The conservation of sea cucumbers in malaysia, taxonomy, ecology and trade*. Herriot Watt Univ. Scotland. pp. 49–63.

Banluta-Batoy C., Alino P.M. & Pocsidio, G.N. 1998 Reproductive development of *Holothuria pulla* and *Holothuria coluber* (Holothuroidea, Echinodermata) in Pamilacan Island, Central Philippines. *Asian Fisheries Society*, 11: 169–176.

Barraza, J. E. & Hasbún, C.R. 2005. Los Equinodermos de El Salvador. *Rev. Biol. Trop. (Int. J. Trop. Biol.)*, Vol. 53(3): 139–146.

Bell, F.J. 1887. Studies in the Holothuroidea - VI. Descriptions of new species. *Proc. Zool. Soc. London*, 531–534.

Brinkhurst, R.O., Linkletter, L.E., Lord, E.I., Conners, S.A. & Dadswell, M.J. 1975. *A preliminary guide to the littoral and sublittoral marine invertebrates of Passamaquoddy Bay*. Special Publication, Huntsman Marine Science Centre, New Brunswick (Canada).

Bruckner, A. 2006. *The proceedings of the Technical Workshop on the conservation of sea cucumbers in the families Holothuriidae and Stichopodidae*. NOAA Technical Memorandum NMFS-OPR 44. Silver Spring, USA. 239 pp.

Byrne, M., Rowe, F. & Uthicke, S. 2010. Molecular taxonomy, phylogeny and evolution in the family Stichopodidae (Aspidochirotida: Holothuroidea) based on COI and 16S mitochondrial DNA. *Mol. Phylogenet. Evol.*, 56: 1068–1081.

Cameron, J.L. & Fankboner, P.V. 1986. Reproductive biology of the commercial sea cucumber *Parastichopus californicus* (Stimpson) (Echinodermata: Holothuroidea). I. Reproductive periodicity and spawning behaviour. *Can. J. Zool.*, 64: 168–175.

Caycedo, I.E. 1978. Holothuroidea (Echinodermata) de aguas someras en la costa de Colombia. *An. Inst. Inv. Mar. Punta Betín*, 10: 149–180.

Chen, J. 2004. Present status and prospects of sea cucumber industry in China. *In* A. Lovatelli, C. Conand, S. Purcell, S. Uthicke, J.-F. Hamel and A. Mercier, eds. *Advances in sea cucumber aquaculture and management,* pp. 25–38. FAO Fisheries Technical Paper No. 463. Rome, FAO. 425 pp.

Cherbonnier, G. 1951. Holothuries de l'Institut royal des Sciences naturelles de Belgique. *Mémoire l'Institut royal des Sciences naturelles de Belgique, deuxième série* 41: 1–65.

Cherbonnier, G. 1955. Résultats Scientifiques des campagnes de la "Calypso". Les Holothuries de la Mer Rouge. *Ann. Inst. Oceanogr.*, Paris, 30: 10–183.

Cherbonnier, G. 1967. Deuxième contribution à l'étude des Holothuries de la mer Rouge collectées par des Israéliens. *Bulletin of the Sea Fisheries Research Station, Haifa* 43: 55–68.

Cherbonnier, G. 1979. Description d'*Actinopyga flammea* nov. sp., et données nouvelles sur deux espèces connues d'Holothuries Aspidochirotes (Échinodermes). *Bull. Mus. natn. Hist. nat.*, (4)1, sect. A 1: 3–12.

Cherbonnier, G. 1980. Holothuries de Novelle-Calédonie. *Bull. Mus. natn. Hist. nat.*, 3: 615–667.

Cherbonnier, G. 1988. Echinodermes: Holothurides. *Faune de Madagascar*, 70: 1–192.

Cherbonnier, G. & Feral, J.-P. 1984. Les Holothuries de Nouvelle-Calédonie: Deuxième contribution. *Bull. Mus. natn. Hist. nat.*, 3: 659–700.

Choo, P.S. 2008a. Population status, fisheries and trade of sea cucumbers in Asia. *In* V. Toral-Granda, A. Lovatelli & M. Vasconcellos, eds. *Sea cucumbers. A global review of fisheries and trade,* pp. 81–118. FAO Fisheries and Aquaculture Technical Paper No. 516. Rome, FAO. 317 pp.

Choo, P.S. 2008b. The Philippines: a hotspot of sea cucumber fisheries in Asia. *In* V. Toral-Granda, A. Lovatelli & M. Vasconcellos, eds. *Sea cucumbers. A global review of fisheries and trade,* pp. 119–140. FAO Fisheries and Aquaculture Technical Paper No. 516. Rome, FAO. 317 pp.

Clark, H.L. 1913. Echinoderms from lower California with description of new species. *Bull. Am. Mus. Nat. Hist.* 32: 185–236.

Clark, H.L. 1922. The holothurians of the genus *Stichopus. Bull. Mus. Comp. Zool. Harv. Univ.*, 65(3): 40–74.

Clark, A.M. & Rowe, F.W.E. 1971. Holothuroidea. *In: Monograph of shallow-water Indo-West Pacific echinoderms*, pp. 171–210. London, Trustees of the British Museum.

Conand, C. 1979. Bêche-de-mer in New Caledonia: weight loss and shrinkage during processing in three species of holothurians. *S.P.C. Fisheries Newsl.*, 19: 14–17.

Conand, C. 1981. Sexual cycle of three commercially important holothurian species (Echinodermata) from the lagoon of New Caledonia. *Bull. Mar. Sci.*, 31(3): 523–544.

Conand, C. 1982. Reproductive cycle and biometric relations in a population of *Actinopyga echinites* (Echinodermata: Holothuroidea) from the lagoon of New Caledonia, western tropical Pacific. *In* J.M. Lawrence, ed. International Echinoderms Conference, Tampa Bay.

Conand, C. 1989. Les holothuries aspidochirotes du lagon de Nouvelle-Calédonie: Biologie, Ecologie et exploitation. Etudes et thèses, ORSTOM: 393 pp.

Conand, C. 1993a. Reproductive biology of the characteristic holothurians from the major communities of the New Caledonia lagoon. *Mar. Biol.*, 116: 439–450.

Conand, C. 1993b. Ecology and reproductive biology of *Stichopus variegatus* an Indo-Pacific coral reef sea cucumber (Echinodermata: Holothuroidea). *Bull. Mar. Sci.*, 52, 3

Conand, C. 1998. Holothurians. *In* K. Carpenter & V. Niem, eds. *The living marine resources of the Western Central Pacific. Vol 2 Cephalopods, crustaceans, holothurians and sharks*, pp. 1157–1190. FAO species identification guide for fishery purposes. Rome, FAO. pp. 687–1396.

Conand, C. 1999. *Manuel de qualité des holothuries commerciales du Sud-Ouest de l'Océan Indien.* PRE/COI: 39 pp.

Conand, C. 2006. Sea cucumber biology: taxonomy; distribution; biology; conservation status. *In* A.W. Bruckner, ed. *The Proceedings of the CITES workshop on the conservation of sea cucumbers in the families Holothuriidae and Stichopodidae*, pp. 33–50. NOAA Technical Memorandum NMFS-OPR-34. USA. 244 pp.

Conand, C. 2008. Population status, fisheries and trade of sea cucumbers in Africa and the Indian Ocean. *In* V. Toral-Granda, A. Lovatelli & M. Vasconcellos, eds. *Sea cucumbers. A global review of fisheries and trade*, pp. 143–193. FAO Fisheries and Aquaculture Technical Paper No. 516. Rome, FAO. 317 pp.

Conand, C. & Frouin, P. 2007. Sea cucumbers in La Reunion. *In* C. Conand & N.A. Muthiga, eds. *Commercial sea cucumbers: a review for the Western Indian Ocean*, pp. 21–29. WIOMSA Book Series No. 5. 66 pp.

Conand, C. & Muthiga, N.A. eds. 2007. *Commercial sea cucumbers: a review for the Western Indian Ocean.* WIOMSA Book Series No 5. 66 pp.

Conand, C., Uthicke, S. & Hoareau, T. 2002. Sexual and asexual reproduction of the holothurian *Stichopus chloronotus* (Echinodermata): a comparison between La Réunion (Indian Ocean) and east Australia (Pacific Ocean). *Invert. Reprod. Develop.*, 41(1–3): 235–242.

Conand, C., Armand, J., Dijoux, N., & Garryer, J. 1998. Fission in a population of *Stichopus chloronotus* on Reunion Island, Indian Ocean. *SPC Beche-de-mer Inf. Bull.*, 10: 15–23.

Conand, C., Muthiga N., Aumeeruddy, R., De La Torre Castro, M., Frouin, P., Mgaya, Y., Mirault, E., Ochiewo, J. & Rasolofonirina, R. 2006. A three-year project on sea cucumbers in the southwest Indian Ocean. National and regional analyses to improve management. A three-year project on sea cucumbers in the southwestern Indian. *SPC Beche-de-mer Inf. Bull.*, 23: 11–15.

Conand, C., Michonneau, F., Paulay, G. & Bruggemann, H. 2010. Diversity of the holothuroid fauna (Echinodermata) in La Réunion (Western Indian Ocean). *Western Indian Ocean J. Mar. Sc.*, 9(2): 145–151.

Cutress, B.M. 1996. Changes in dermal ossicles during somatic growth in Caribbean littoral sea cucumbers (Echinoidea: Holothuroidea: Aspidochirotida). *Bull. Mar. Sci.*, 58: 44–116.

Deichmann, E. 1940a. The holothurioidea collected by the Velero III during the years 1932 to 1939. Port. I, Dendrochirota. *Allan Hancock Pacif. Exped.*, 8(3): 61–194.

Deichman, E. 1940b. Report on the holothurian collected by Harvard-Havana Expedition "Atlantics" 1938-1939, with review on Molpadonia of the Atlantic Ocean Soc. *Cubana Hist. Nat.*, 14(3): 183–240.

Deichmann, E. 1957. The Littoral Holothurians of the Bahama Islands. *American Museum Novitates* 1821: 1–20.

Deichman, E. 1958. *The Holothurioidea collected by The Velero II and IV during the years of 1932 to 1954; Part II Aspidochirotida.* The University of Southern California Publications, Allan Hancock Pacific Expeditions, Volume II, Number 2. Los Angeles, CA.

Dendy, A. 1897. Observations on the Holothurians of New Zealand; with Descriptions of four New Species, and an Appendix on the Development of the Wheels in Chiridota. *J. Linn. Soc. London Zool.* 26: 22–52, pls 3–7.

Dendy, A., & Hindle, E. 1907. Some additions to our knowledge of the New Zealand holothurians. *J. Linn. Soc. London Zool.*, 30: 95–125.

DFO. 2002. Giant Red Sea Cucumber. DFO Science Stock Status Report C6-10. http://www.pac.dfo-mpo.gc.ca/sci/psarc/SSRs/Invert/C6-10.pdf

Eeckhaut, I., Parmentiers, E., Becker, P., Gomez da Silva, S. & Jangoux, M. 2004. Parasites and biotic diseases in field and cultivated sea cucumbers. *In* A. Lovatelli, C. Conand, S. Purcell, S. Uthicke, J.-F. Hamel and A. Mercier, eds. *Advances in sea cucumber aquaculture and management*, pp. 311–326. FAO Fisheries Technical Paper No. 463. Rome, FAO. 425 pp.

Engstrom, N.A. 1980. Reproductive cycles of *Holothuria* (Halodeima) *floridana, H.* (H.) *mexicana* and their hybrids (Echinodermata: Holothuroidea) in southern Florida, USA. *Int. J. Invertebr. Reprod.*, 2: 237–244.

Fajardo-Leon, M.C., Suárez-Higuera, MCL, del Valle-Manriquez,A. & Hernandez-López, A. 2008. Reproductive biology of the sea cucumber *Parastichopus parvimensis* (Echinodermata: Holothuroidea) at Isla Natividad and Bahía Tortugas, Baja California Sur, México. *Ciencias Marinas*, 34 (Abstract).

Falk-Petersen, I.B. 1982. Breeding season and egg morphology of echinoderms in Balsfjorden, Northern Norway. *Sarsia*, 67: 215–221.

Féral, J-P. & Cherbonnier G. 1986. Les holothurides. *In* A.Guille, P. Laboute, J.-L, Menou, eds. *Guide des étoiles de mer, oursins et autres echinoderms du lagon de Nouvelle-Calédonie*, pp. 54–107. Paris, Editions de l'ORSTOM.

Foglietta, L.M., Camejo, M.I., Gallardo, L. & Herrera, F.C. 2004. A maturity index of holothurians exhibiting asynchronous development of gonad tubules. *J. Exp. Mar. Biol. Ecol.*, 300: 19–30.

Friedman, K., Ropeti, E. & Tafileichig, A. 2008. Development of a management plan for Yap's sea cucumber fishery. *SPC Beche-de-mer Inf. Bull.*, 28: 7–13.

Friedman, K., Pakoa, K., Kronen, M., Chapman, L., Sauni, S., Vigliola, L., Boblin, P. & Magron, F. 2008. *Vanuatu country report: Profile and results from survey work at Paunangisu village, Moso Island, Uri and Uripiv islands and the Maskelyne Archipelago (July to December 2003).* SPC PROCFish/C/CoFish, Noumea, 182 pp.

Friedman, K., Pinca, S., Kronen, M., Boblin, P., Chapman, L., Magron, F., Vunisea, A. & Labrosse, P. 2009. *Tonga country report: profiles and results from survey work at Ha'atafu, Manuka, Koulo and Lofanga (November and December 2001; March to June 2002; April to June, September and October 2008).* SPC PROCFish/C/CoFish, Noumea, 222 pp.

Gabr, H. R., Ahmed, A.I., Hanafy, M.H., Lawrence, A.J., Ahmed, M.I. & El-Etreby, S.G. 2004. Mariculture of sea cucumber in the Red Sea - the Egyptian experience. *In* A. Lovatelli, C. Conand, S. Purcell, S. Uthicke, J.-F. Hamel & A. Mercier, eds. *Advances in sea cucumber aquaculture and management*, pp. 373–384. FAO Fisheries Technical Paper No. 463. Rome, FAO. 425 pp.

Garcia-Arraras, J.E., Torres-Avillan, I. & Ortiz-Miranda, S. 1991. Cells in the intestinal system of holothurians (Echinodermata) express cholecystokinin-like immunoreactivity. *General and Comparative Endocrinology*, 83(2): 233–242.

Gosliner, T., Behrens, D. & Williams, G. 1996. *Coral reef animals of the Indo-Pacific*. Monterey, USA, Sea Challengers.

Graaf, de M., Geertjes, G.J. & Videler, J.J. 1999. Observations on spawning of scleractinian corals and other invertebrates on the reefs of Bonaire (Netherlands Antilles, Caribbean). *Bull. Mar. Sci.*, 64: 189–194.

Graham, J.C.H. & Battaglene, S.C. 2004. Periodic movement and sheltering behaviour of *Actinopyga mauritiana* (Holothuroidea:Aspidochirotidae) in Solomon Islands. *SPC Beche-de-mer Inf. Bull.*, 19: 23–31.

Gudimova, E.N. & Antsiferova A.V. 2006. Gonadal morphology and oogenetic stages of *Cucumaria frondosa* from the Barents Sea: comparative aspect. Abstract book of 12th International Echinoderm Conference, University of New Hampshire, Durham, NH, August, 7-11, pp. 79–80.

Guzmán, H.M. & Guevara, C.A. 2002. Population structure, distribution and abundance of three commercial species of sea cucumber (Echinodermata) in Panama. *Caribbean J. Sci.*, Vol. 38, No. 3–4: 230–238.

Guzmán, H.M., Guevara, C.A. & Hernández, I.C. 2003. Reproductive cycle of two commercial species of sea cucumber (Echinodermata: Holothuroidea) from Caribbean Panama. *Mar. Biol.*, 142: 271–279.

Hamel, J.-F. & Mercier, A. 1995. Spawning of the sea cucumber *Cucumaria frondosa* in the St. Lawrence Estuary, eastern Canada. *SPC Beche-de-mer Inf. Bull.*, 7: 12–18.

Hamel, J.-F. & Mercier, A. 1996a. Evidence of chemical communication during the gametogenesis of holothuroids. *Ecology*, 77: 1600–1616.

Hamel, J.-F. & Mercier, A. 1996b. Early development, settlement, growth, and spatial distribution of the sea cucumber *Cucumaria frondosa* (Echinodermata: Holothuroidea). *Can. J. Fish. Aquat. Sci.*, 53: 253–271.

Hamel, J.-F. & Mercier, A. 1996c. Studies on the reproductive biology of Atlantic sea cucumber *Cucumaria frondosa*. *SPC Beche-de-mer Inf. Bull.*, 8: 22–33.

Hamel, J.-F. & Mercier, A. 1996d. Gonad morphology and gametogenesis of the sea cucumber *Cucumaria frondosa*. *SPC Beche-de-mer Inf. Bull.*, 8: 22–33.

Hamel, J.-F. & Mercier, A. 1996e. Gamete dispersion and fertilisation success of the sea cucumber *Cucumaria frondosa*. *SPC Beche-de-mer Inf. Bull.*, 8: 34–40.

Hamel, J.-F. & Mercier, A. 2008. Population status, fisheries and trade of sea cucumbers in temperate areas of the Northern Hemisphere. *In* M.V. Toral-Granda, A. Lovatelli & M. Vasconcellos, eds. *Sea cucumbers. A global review on fisheries and trade*, pp. 257–291. FAO Fisheries and Aquaculture Technical Paper No. 516. Rome, FAO. 317 pp.

Hamel, J.-F., Conand, C., Pawson, D.L. & Mercier, A. 2001. The sea cucumber *Holothuria scabra* (Holothuroidea: Echinodermata): its biology and exploitation as beche-de-mer. *Advances in Marine Biology*, 41: 129–223.

Hamel, J.-F., Ycaza-Hidalgo, R. & Mercier, A. 2003. Larval development and juvenile growth of the Galapagos sea cucumber *Isostichopus fuscus*. *SPC Beche-de-mer Inf. Bull.*, 18: 3–8.

Harbo, R.M. 1999. *Whelks to Whales, Coastal Marine Life of the Pacific Northwest*. Madeira Park, Canada, Harbour Publishing.

Harriott, V.J. 1985. Reproductive biology of three congeneric sea cucumber species, *Holothuria atra, H. impatiens* and *H. edulis*, at Heron Reef, Great Barrier Reef. Aust. *J. Mar. Freshwat. Res.*, 36: 51–57.

Hasbun, C.R. & Lawrence, A.J. 2002. An annotated description of shallow water holothurians (Echinodermata: Holothuroidea) from Cayos Cochinos, Honduras. *Rev. Biol. Trop.*, 50 (2): 669–678.

Hearn, A. & Pinillos, F. 2006. Baseline information on the warty sea cucumber *Stichopus horrens* in Santa Cruz, Galapagos, prior to the commencement of an illegal fishery. *SPC Beche-de-mer Inf. Bull.*, 24: 3–10.

Hearn, A., Mora, J. & Moreno, J. 2008. *Evaluación del pepino de mar Stichopus horrens en la Reserva Marina de Galápagos*. Informe 2007. Santa Cruz, Galapagos, Ecuador, Fundación Charles Darwin. 35 pp.

Heding, S.G. & Panning, A. 1954. Phyllophoridae. Eine Bearbeitung der polytentaculaten dendrochiroten Holothurien des Zoologische Museum in Kopenhagen. *Spolia Zoologica Musei Hauniensis* 13: 1–209.

Hendler, G., Miller, J.E., Pawson, D.L. & Kier, P.M. 1995. *Sea stars, sea urchins and allies. Echinoderms of Florida and the Caribbean*. Washington, Smithsonian Institution Press, i-xi, 1–390.

Herrero Pérezrul, M.D., Reyes Bonilla, H., García Domínguez, F. & Cintra Buenrostro, C.E. 1999. Reproduction and growth of *Isostichopus fuscus* (Echinodermata: Holothuroidea) in the southern Gulf of California, México. *Mar. Biol.*, 135: 521–532.

Hickman, C.J. 1998. *A field guide to sea stars and other echinoderms of Galapagos*. Lexington, USA, Sugar Spring Press. 83 pp.

Hill, N. 2008. Information on the sea cucumber fishery in the Querimbas Archipelago, Mozambique. *SPC Beche-de-mer Inf. Bull.*, 27: 16.

Hoareau, T. & Conand, C. 2002. Sexual reproduction of *Stichopus chloronotus*, a fissiparous sea cucumber, on Reunion Island, Indian Ocean. *SPC Beche-de-mer Inf. Bull.*, 15: 4–12.

Hooker, Y., Solís-Marín, F.A. & Lleellish, M. 2005. Equinodermos de las Islas Lobos de Afuera (Lambayeque, Perú). *Rev. peru. biol.*, 12(1): 77–82.

Hopper, D.R., Hunter, C.L. & R.H. Richmond. 1998. Sexual reproduction of the tropical sea cucumber *Actinopyga mauritania* (Echinodermata: Holothuroidea) in Guam. *Bull. Mar. Sci.*, 63(1):1–9.

IMARPE. Undated. "Pepino de Mar" *Athyonidium chilensis* (Semper, 1868). http://www.imarpe.gob.pe/imarpe/index.php?id_detalle=00000000000000007869. Consulted on March 2010.

James, D.B. 1989. Echinoderms of Lakshadweep and their zoogeography. *Bull. Cent. Mar. Fish. Res. Inst.*, 43: 97–144.

James, D.B. 1994. Holothurians resource from India and their exploitation. *Bulletin of the Central Marine Fisheries Research Institute*, 46: 27–31.

James, D.B. & James, P.S.B.R. 1994. Handbook on Indian sea cucumbers. *CMFRI Spl. Publ.*, No. 59: 1–46.

James, D.B., Gandhi, A.D., Palaniswamy, N. & Rodrigo, J.X. 1994a. Hatchery techniques and culture of the sea-cucumber *Holothuria scabra*. *Central Marine Fisheries Research Institute Special Publication*, 57: 1–40.

James, D.B., Rajapandian, M.E., Gopinathan, C.P. & Baskar, B.K. 1994b. Breakthrough in induced breeding and rearing of the larvae and juveniles of *Holothuria* (*Metriatyla*) *scabra* Jaeger at Tuticorin. *Bulletin of the Central Marine Fisheries Research Institute*, 46: 66–70.

Jordan, A.J. 1972. On the ecology and behaviour of *Cucumaria frondosa* (Echinodermata: Holothuroidea) at Lamoine Beach, Maine. Ph. D. Thesis. Univ. Maine. Orono. 75 pp.

Kinch, J. 2004. *A Review of the Beche-de-mer Fishery and its Management in Papua New Guinea*. A report prepared by the Motupore Island Research Centre for the National Fisheries Authority, Port Moresby, Papua New Guinea, 133 pp.

Kinch, J., Purcell, S., Uthicke, S. & Friedman, K. 2008a. Population status, fisheries and trade of sea cucumbers in the Western Central Pacific. *In* V. Toral-Granda, A. Lovatelli & M. Vasconcellos, eds. *Sea cucumbers. A global review of fisheries and trade*, pp. 7–55. FAO. Fisheries and Aquaculture Technical Paper No. 516. Rome, FAO. 317 pp.

Kinch, J., Purcell, S., Uthicke, S. & Friedman, K. 2008b. Papua New Guinea: a hotspot of sea cucumber fisheries in the Western Central Pacific. *In* V. Toral-Granda, A. Lovatelli & M. Vasconcellos, eds. *Sea cucumbers. A global review of fisheries and trade*, pp. 57–77. FAO. Fisheries and Aquaculture Technical Paper No. 516. Rome, FAO. 317 pp.

Kohtsuka, H., Arai, S. & Uchimura, M. 2005. Observation of asexual reproduction by natural fission of *Stichopus horrens* in Okinawa Islans, Japan. *SPC Beche-de-mer Inf. Bull.*, 22: 23.

Kronen, M., Friedman, K., Pinca, S., Chapman, L., Awira, R., Pakoa, K., Vigliola, L., Boblin, P. & Magron, F. 2008. *French Polynesia country report: Profiles and results from survey work at Fakarava, Maatea, Mataiea, Raivavae and Tikehau (September – October 2003, January – March 2004, April – June 2006)*. SPC PROCFish/C/CoFish, Noumea, 246 pp.

Labe, L. 2009. Sea cucumber fisheries, utilization and trade in the Philippines. *In* SEAFDEC-Secretariat, ed. *Report of the Regional Study on Sea Cucumber Fisheries, Utilization and Trade in Southeast Asia (2007-2008)*, pp. 68–94. Bangkok, SEAFDEC.

Laboy-Nieves, E.N & Conde, J.E. 2006. A new approach for measuring *Holothuria mexicana* and *Isostichopus badionotus* for stock assessments. *SPC Beche-de-mer Inf. Bull.*, 24: 39–44.

Lambert, P. 1984. British Columbia marine faunistic survey report holothurians from the Northeast Pacific. *Canadian Tech. Rep. Fish. Aquat. Sci.*, N. 1234: 1–30.

Lambert, P. 2007. *Sea Cucumbers of British Colombia, Southeast Alaska and Puget Sound*. Royal British Colombia Museum, 166 pp.

Lane, D.J.W. 1999. Distribution and abundance of *Thelenota rubralineata* in the Western Pacific: Some conservation issues. *SPC Beche-de-mer Inf. Bull.*, 11: 19–21.

Lane, D.J.W., Marsh, L.M., VandenSpiegel, D. & Rowe, F.W.E. 2000. Echinoderm fauna of the South China Sea: An inventory and analysis of distribution patterns. *Raffles Bull. Zool., Suppl.*, 8: 457–492.

Latypov, Y.Y. 2001. Communities of coral reefs of Central Vietnam. *Russian Journal of Marine Biology*, Vol. 27, No. 4, 2001, pp. 197–200. [Original Russian Text Copyright © 2001 by Biologiya Morya, Latypov].

Lawrence, A.J., Ahmed, M., Hanafy, M., Gabr, H., Ibrahim, A. & Gab-Alla, A.A-F.A. 2004. Status of the sea cucumber fishery in the Red Sea - the Egyptian experience. *In* A. Lovatelli, C. Conand, S. Purcell, U. Uthicke, J-F. Hamel & A. Mercier, eds *Advances in sea cucumber aquaculture and management*, pp. 79–90. FAO Fisheries Technical Paper No. 463. Rome, FAO. 425 pp.

Laxminarayana, A. 2006. Asexual reproduction by induced transverse fission in the sea cucumbers *Bohadschia marmorata* and *Holothuria atra*. *SPC Beche-de-mer Inf. Bull.*, 23: 35–37

Levin V.S. 1995. Japanese sea cucumber *Cucumaria japonica* in the eastern seas of Russia. *SPC Beche-de-mer Inf. Bull.*, 7: 18–19.

Levin, V.S. & Gudimova, E.N. 2000. Taxonomic interrelations of holothurians *Cucumaria frondosa* and *C. japonica* (Dendrochirotida, Cucumaridae). *SPC Beche-de-mer Inf. Bull.*, 13: 22–29.

Li, X. 2004. Fishery and resource management of tropical sea cucumbers in the islands of the South China Sea. *In* A. Lovatelli, C. Conand, S. Purcell, S. Uthicke, J.-F. Hamel & A. Mercier, eds. *Advances in sea cucumber aquaculture and management*, pp. 261–265. FAO Fisheries Technical Paper No. 463. Rome, FAO. 425 pp.

Liao, Y. 1980. The Aspidochirote Holothurians of China with erection of a new genus. *In* M. Jangoux ed. *Proceedings of European Colloquium on Echinoderms: Present and past*, pp. 115–120. Rotterdam, Netherlands, A.A. Balkema Publishers.

Lovatelli, A., Conand, C., Purcell, S.; Uthicke, S., Hamel, J.-F.; Mercier, A. eds. 2004. *Advances in sea cucumber aquaculture and management*. FAO Fisheries Technical Paper. No. 463. Rome, FAO. 425 pp.

Ludwig, H.L. 1875. Beiträge zur Kenntniss der Holothurien. Arbeiten aus dem Zoologischen zootom. Institut in Würzburg 2(2): 77–120, pl. 6–7.

Maluf, L.Y. 1991. Echinoderms of the Galapagos Islands. *In* M.J. James, ed. *Galapagos Marine Invertebrates*, pp. 345–367. New York & London, Plenum Press.

Massin, C. 1996. Results of the Rumphius Biohistorical Expedition to Ambon (1990). Part 4. The Holothuroidea (Echinodermata) collected at Ambon during the Rumphius Biohistorical Expedition. *Zoologische Verhandelingen* 307: 1–53.

Massin, C. 1999. Reef-dwelling Holothuroidea (Echinodermata) of the Spermonde Archipelago (South-West Sulawesi, Indonesia). *Zoologische Verhandelingen*, 329: 3–144.

Massin, C. 2007. Redescription of *Stichopus naso* Semper, 1868 (Holothuroidea, Aspidochirotida, Stichopodidae). *Bull. de l'Institut Royal des Sciences Naturelles de Belgique*, 77: 123–130.

Massin, C. & Lane, D. 1991. Description of a new species of sea cucumber (Stichopodidae, Holothuroidea, Echinodermata) from the Eastern Indo-Malayan Archipelago: *Thelenota rubralineata* n. sp. *Micronesica*, 24(1): 57–64.

Massin, C., Mercier, A. & Hamel, J.-F. 2000. Ossicle change in *Holothuria scabra* with a discussion of ossicle evolution within the Holothuriidae (Echinodermata). *Acta Zoologica*, 81: 77–91.

Massin, C., Rasolofonirina, R., Conand, C. & Samyn, Y. 1999. A new species of *Bohadschia* (Echinodermata, Holothuroidea) from the Western Indian Ocean with a redescription of *Bohadschia subrubra. Bull. de l'Institut Royal des Sciences Naturelles de Belgique*, 69: 151–160.

Massin, C., Zulfigar, Y., Tan Shau Hwai, A. & Rizal Boss, S.Z. 2002. The genus *Stichopus* (Echinodermata: Holothuroidea) from the Johore Marine Park (Malaysia) with the description of two new species. *Bulletin de l'Institut royal des Sciences naturelles de Belgique*, 72: 73–99.

Massin, C., Uthicke, S., Purcell, S.W., Rowe, F.W.E. & Samyn, Y. 2009. Taxonomy of the heavily exploited Indo-Pacific sandfish complex (Echinodermata: Holothuriidae). *Zoological Journal of the Linnean Society*, 155: 40–59.

Mercier, A., Battaglene, S.C. & Hamel, J.-F. 1999. Daily burrowing cycle and feeding activity of juvenile sea cucumbers *Holothuria scabra* in response to environmental factors. *J. Exp. Mar. Biol. Ecol.*, 239:125–156.

Mgaya, Y. & Mmbaga, T. 2007. Sea cucumbers in Tanzania. *In* C. Conand & N.A. Muthiga, eds. *Commercial sea cucumbers: a review for the Western Indian Ocean.* WIOMSA Book Series, No. 5 v + 66 pp.

Miller, J.E. & Pawson, D.L. 1984. Holothurians (Echinodermata: Holothuroidea). *Memoirs of the Hourglass Cruises* 7: 1–79

Moraes, G., Norhcote, P.C., Kalinin, V.I., Avilov, S.A., Silchenko, A.S., Dmitrenok, P.S., Stonik, V.A. & Levin, V.S. 2004. Structure of the major triterpene glycoside from the sea cucumber *Stichopus mollis* and evidence to reclassify this species into the new genus *Australostichopus*. *Biochem. Syst. Ecol.*, 32: 637–650.

Mokretsova, N.D. & Koshkaryova, L.N. 1983. Some data about the distribution, ecology and reproduction biology of *Cucumaria japonica* (Semper). *In*: *Mariculture in the Far East*, pp. 46–51. Vladivostok.

Mosher, C. 1980. Distribution of *Holothuria arenicola* Semper in the Bahamas with observations on habitat, behaviour and feeding activity (Echinodermata: Holothuroidea). *Bulletin of Marine Science*, 30(1): 1–12.

Mosher, C. 1982. Spawning behaviour of the aspidochirote holothurian *Holothuria mexicana* Ludwig. *In* J.M. Lawrence, ed. *Proc. Int. Echinoderm. Conf. Tampa Bay*, pp. 467–468. Rotterdam, Netherlands, A.A. Balkema Publishers.

Muthiga, N., Ochiewo, J. & Kawaka, J. 2007. Sea cucumbers in Kenya. *In* C. Conand & N.A. Muthiga, eds. *Commercial sea cucumbers: a review for the Western Indian Ocean.* WIOMSA Book Series No. 5 v + 66 pp.

Muthiga, N., Ochiewo, J. & Kawaka, J. 2010. Strengthening capacity to sustainably manage sea cucumber fisheries in the western Indian Ocean. *SPC Beche-de-mer Inf. Bull.*, 30: 3–9.

Naidenko, V.P. & Levin, V.S. 1983. Rearing of the holothurian *Cucumaria japonica* in the laboratory. *Biol. Morya (Vladivost.)*, 61–65.

Nguyen An Khang. 2006. Echinoderm biodiversity of Coral reef ecosystem in Phu Quoc, 5/2006. Scientific report of project (unpublished) for UNEP/GEF Project: "Reversing Environmental Degradation Trends in the South China Sea and Gulf of Thailand" UNEP/GEF/SCS/RWG-CR.6/7. Original Vietnamese Text, translated by Nguyen An Khang, Nha Trang Institute of Oceanography, Viet Nam.

Nguyen Huu Phung. 1994. Surveying special marine products of coastal zones of mainland and islands in Vietnam (unpublished). Scientific report of project, PGS.PTS. Original Vietnamese Text, translated by Nguyen An Khang, Nha Trang Institute of Oceanography, Viet Nam.

Oganesyan, S.A. & Grigorjev, G.V. 1998. Reproductive cycle of the holothuroid *Cucumaria frondosa* in the Barents Sea. Echinoderms: San Francisco, pp. 489–492. Rotterdam, Netherlands, A.A. Balkema Publishers.

Otero-Villanueva, M.M. 2005. *Stock assessment of sea cucumbers from Phu Quoc.* Scientific report of project (unpublished) for Can Tho University, Viet Nam. 55 pp.

Otero-Villanueva, M.M. & Ut, V.N. 2007. Sea cucumber fisheries around Phu Quoc Archipelago: A cross-border issue between South Vietnam and Cambodia. *SPC Beche-de-mer Inf. Bull.*, 25: 32–36.

Panning A. 1929. Die Gattung Holothuria. (1. Teil).- *Mitteilungen aus dem Zoologischen Staatsinstituut und Zoologischen Museum in Hamburg* 44: 91–138.

Panning, A. 1944. Die Trepangfisherei. *Mitteilungen aus dem Hamburgischen Zoologischen Museum und Institut*, 49: 2–76.

Pawson, D.L. 1969. Holothuroidea from Chile report No. 46 of the Lund University Chile Expedition 1948-1949. *Sarsia*, 38: 121–145.

Pawson, D.L. 1970. The Marine Fauna of New Zealand: Sea Cucumbers (Echinodermata: Holothuroidea). *Bull. N.Z. Dep. Scient. Ind. Res.*, 201.

Pawson, D.L. 1986. Phylum Echinodermata. 522–541. *In* W. Sterrer, ed. *Marine fauna and flora of Bermuda: a systematic guide to the identification of marine organisms.* John Wiley and Sons, New York.

Pires-Nogueira, R.L., Freire, C.B., Rezende, C.R. & Campos-Creasey, L.S. 2001. Reproductive cycle of the sea cucumber *Isostichopus badionotus* (Echinodermata) from Baía da Ilha Grande, Rio de Janeiro, Brazil (abstract). *In: Abstracts of the 9th international congress of invertebrate reproduction and development.* Rhodes University, Grahamstown, South Africa, 15–20 July 2001.

Purcell, S.W. 2010. *Managing sea cucumber fisheries with an ecosystem approach.* Edited/compiled by Lovatelli, A.; M. Vasconcellos and Y. Yimin. FAO Fisheries and Aquaculture Technical Paper No. 520. Rome, FAO. 157 pp.

Purcell, S.W. & Tekanene, M. 2006. Ontogenetic changes in colouration and morphology of white teatfish, *Holothuria fuscogilva*, juveniles at Kiribati. *SPC Beche-de-mer Inf. Bull.*, 23: 29–31.

Purcell, S.W., Gossuin, H. & Agudo, N.S. 2009. *Status and management of the sea cucumber fishery of La Grande Terre, New Caledonia.* WorldFish Center Studies and Review N° 1901. Penang, Malaysia, The WorldFish Center. 136 pp.

Purcell, S.W., Tardy, E., Desurmont, A. & Friedman, K.J. 2008. Commercial holothurians of the tropical Pacific [Poster]. Noumea, New Caledonia, Secretariat of the Pacific Community.

Rasolofonirina, R. 2007. Sea cucumbers in Madagascar. *In* C. Conand & N.A. Muthiga, eds. *Commercial sea cucumbers: a review for the Western Indian Ocean.* WIOMSA Book Series No. 5 v + 66 pp.

Reyes Bonilla, H. & Herrero Pérezrul, M.D. 2003. Population parameters of an exploited population of *Isostichopus fuscus* (Holothuroidea) in the southern Gulf of California, Mexico. *Fisheries Research*, 59: 423–430.

Rodríguez-Milliet, E. & Pauls, S.M. 1998. Sea cucumbers fisheries in Venezuela. *Proc. 9th Inter. Echinoderm Conf.*, 513–516.

Rowe, F.W.E. & Doty, J.E. 1977. The shallow-water Holothurians of Guam. *Micronesica*, 13(2): 217–250.

Samyn, Y. 2003. Shallow-water Holothuroidea (Echinodermata) from Kenya and Pemba Island (Tanzania). *Studies in Afrotropical Zoology*, 292: 1–158.

Samyn, Y. & Massin, C. 2003. The holothurian subgenus *Mertensiothuria* (Aspidochirotida: Holothuriidae) revisited. *J. Nat. Hist.*, 37 (20): 2487–2519.

Samyn, Y., VandenSpiegel, D. & Massin, C. 2005. Sea Cucumbers of the Comoros Archipelago. *SPC Bêche-de-mer Inf. Bull.*, 22: 14–19.

Samyn, Y., VandenSpiegel, D. & Massin, C. 2006a. Taxonomie des holothuries des Comores. *AbcTaxa*, Vol 1, i–iii, 130 pp.

Samyn, Y., VandenSpiegel, D. & Massin, C. 2006b. A new Indo-West Pacific species of *Actinopyga* (Holothuroidea: Aspidochirotida: Hotothuriidae). *Zootaxa*, 1138: 53–68.

Samyn, Y., VandenSpiegel, D., Franklin, A., Réveillon, A., Segers, H. & van Goethem, J. 2004. The Belgian focal point to the global taxonomy initiative and its role in strengthening individual and institutional taxonomic capacity for, inter alia, sea cucumbers. *In* A. Lovatelli, C. Conand, S. Purcell, S. Uthicke, J.-F. Hamel & A. Mercier, eds. *Advances in sea cucumber aquaculture and management*, pp. 415-418. FAO Fisheries Technical Paper. No. 463. Rome, FAO. 425 pp.

Saveljeva, T.S. 1941. On the fauna of holothurians of the Far East seas. *Invest. of the USSR seas*, 1: 73–103.

Sewell, M.A. 1987. The reproductive biology of *Stichopus mollis* (Hutton). M.Sc. thesis, Univ. of Auckland, New Zealand. 90 pp.

Sewell, M.A. 1990. Aspects of the ecology of *Stichopus mollis* (Echinodermata: Holothuroidea) in north-eastern New Zealand. *N. Z. J. Mar. Freshw. Res.*, 24: 87–93.

Sewell, M.A. 1992 Reproduction of the temperate aspidochirote *Stichopus mollis* (Echinodermata: Holothuroidea) in New Zealand. *Ophelia*, 35: 103–121.

Sewell, M.A. & Bergquist, P.R. 1990. Variability in the reproductive cycle of *Stichopus mollis* (Echinodermata: Holothuroidea). *Invert. Reprod. Develop.*, 17: 1–7.

Shelley, C.C. 1985. Growth of *Actinopyga echinites* and *Holothuria scabra* (Holothurioidea: Echinodermata) in Papua New Guinea. *In* C. Gabrié *et al.* eds. *Proceedings of the 5th International Coral Reef Congress*, pp. 297–330. Moorea, French Polynesia, Antenne Museum, Ecole Pratique des Hautes Etudes.

Shiell, G.R. 2004. Density of *Holothuria nobilis* and distribution patterns of common holothurians on coral reefs of Northwestern Australia. *In* A. Lovatelli, C. Conand, S. Purcell, S. Uthicke, J.-F. Hamel & A. Mercier, eds. *Advances in sea cucumber aquaculture and management*, pp. 231–237. FAO Fisheries Technical Paper No. 463. Rome, FAO. 425 pp.

Shiell, G.R. 2007. Spatial distribution and temporal shifts in the biology of the commercial sea cucumber *Holothuria whitmaei* [Echinodermata: Holothuroidea], Ningaloo Reef, Western Australia. *SPC Beche-de-mer Inf. Bull.*, 26: 35–36.

Shiell, G.R. & Uthicke, S. 2005. Reproduction of the commercial sea cucumber *Holothuria whitmaei* (Holothuroidea: Aspidochirotida) in the Indian and Pacific Ocean regions of Australia. *Marine Biology*, 148: 973–986.

Skewes, T., Kinch, J., Polon, P., Dennis, D., Seeto, P., Taranto, T., Lokani, P., Wassenberg, T., Koutsoukos, A. & Sarke, J. 2002. *Research for the sustainable use of beche-de-mer resources in the Milne Bay Province, Papua New Guinea.* CSIRO Division of Marine Research Final Report. 40 pp.

Sluiter, C.P. 1910. Westindische Holothurien. *Zool. Jahrb. Jena Suppl.* 11: 331–342.

SPC & NFA. 2003. Papua New Guinea sea cucumber and beche-de-mer identification cards.

Solís-Marín, F.A., Arriaga-Ochoa, J.A., Laguarda-Figueras, A., Frontana-Uribe, S.C. & Duran-Gonzalez, A. 2009. *Holoturoideos (Echinodermata: Holothuroidea) del Golfo de California.* Distrito Federal, Mexico, Jimenez Editores e Impresores. 177 pp.

Tehranifard, A., Uryan, S., Vosoghi, G., Mohamadreza, S. & Nikoyan, A. 2006. Reproductive cycle of *Stichopus herrmanni* from Kish Island, Iran. *SPC Beche-de-mer Inf. Bull.,* 24: 22–27.

Theel, H. 1886. Report on the Holothuroidea. Part II. *Rep. Scient. Results Voy. Challenger Zool.,* 14: 1–290.

Toral-Granda, M.V. 2005. The use of calcareous ossicles for the identification of the Galápagos sea cucumber *Isostichopus fuscus* in the international market. *SPC Beche-de-mer Inf. Bull.,* 22: 3–5.

Toral-Granda, M.V. 2007. *The biological and trade status of sea cucumbers in the families Holothuriidae and Stichopodidae.* CoP14 Doc 62. The Hague, June 2007. http://www.cites.org/eng/com/AC/22/index.shtml

Toral-Granda, M.V. 2008a. Population status, fisheries and trade of sea cucumbers in Mexico, Central and South America. *In* M.V. Toral-Granda, A. Lovatelli & M. Vasconcellos, eds. *Sea cucumbers. A global review on fisheries and trade,* pp. 213–229. FAO Fisheries and Aquaculture Technical Paper No. 516. Rome, FAO. 317 pp.

Toral-Granda, M.V. 2008b. Galapagos Islands: a hotspot of sea cucumber fisheries in Mexico, Central and South America. *In* M.V. Toral-Granda, A. Lovatelli & M. Vasconcellos, eds. *Sea cucumbers. A global review on fisheries and trade,* pp. 231–253. FAO Fisheries and Aquaculture Technical Paper No. 516. Rome, FAO. 317 pp.

Toral-Granda, M.V. & Martínez, P. 2007. Reproductive biology and population structure of the sea cucumber *Isostichopus fuscus* (Ludwig, 1875) (Holothuroidea) in Caamaño, Galápagos Islands, Ecuador. *Marine Biology,* 151: 2091–2098.

Toral-Granda, M.V., Lovatelli, A. & Vasconcellos, M., eds. 2008. *Sea cucumbers. A global review on fisheries and trade.* FAO Fisheries and Aquaculture Technical Paper No. 516. Rome, FAO. 2008. 317 pp.

Uthicke, S. 1997. Seasonality of asexual reproduction in *Holothuria* (*Halodeima*) *atra, H.* (*H.*) *edulis* and *Stichopus chloronotus* (Holothuroidea: Aspidochirotida) on the Great Barrier Reef. *Mar. Biol.,* 129: 435–441.

Uthicke, S. 2004. Overfishing of holothurians: lessons from the Great Barrier Reef. Advances in sea cucumber aquaculture and management. *In* A. Lovatelli, C. Conand, S. Purcell, S. Uthicke, J.-F. Hamel & A. Mercier, eds. pp. 163–171. FAO Fisheries Technical Paper No. 463. Rome, FAO. 425 pp.

Uthicke, S. & Benzie, J. 2000. The effect of bêche-de-mer fishing on densities and size structure of *Holothuria nobilis* (Echinodermata: Holothurioidea) populations on the Great Barrier Reef. *Coral Reefs,* 19: 271–276.

Uthicke, S & Conand, C. 2005. Local examples of beche-de-mer overfishing: An initial summary and request for information. *SPC Beche-de-mer Inf. Bull.,* 21: 9–14.

Uthicke, S., O'Hara, T.D. & Byrne, M. 2004. Species composition and molecular phylogeny of the Indo-Pacific teatfish (Echinodermata: Holothuroidea) bêche-de-mer fishery. *Marine and Freshwater Research,* 55: 1–12.

Uthicke, S., Purcell, S. & Blockmans, B. 2005. Natural hybridization does not dissolve species boundaries in commercially important sea cucumbers. *Biological Journal of the Linnean Society,* 85: 261–270.

Uthicke, S., Choo, P.S. & Conand, C. 2008. Observations of two Holothuria species (*H. theeli* and *H. portovallartensis*) from the Galapagos. *SPC Beche-de-mer Inf. Bull.,* 27: 4.

Uthicke, S., Byrne, M., & Conand, C. 2010. Genetic barcoding of commercial Beche-de-mer species (Echinodermata: Holothuroidea). *Molecular Ecology Resources,* 10: 634–646.

van Veghel, M.L.J. 1993. Multiple species spawning on Curacao reefs. *Bull. Mar. Sci.,* 52: 1017–1021.

Wiedemeyer, W.L. 1994. Biology of small juveniles of the tropical holothurian *Actinopyga echinites*: growth, mortality and habitat preferences. *Mar. Biol.,* 120: 81–93.

EXPLANATION OF THE SYSTEM

Italics : Valid scientific names (double entry by genera and species)

Italics : Synonyms and misidentifications

ROMAN : Family names

ROMAN : Names of orders, class, subclass, superfamilies

Roman : FAO names

Roman : Local names

PLATE I	1. *Actinopyga agassizii* (live)
	2. *Actinopyga echinites* Pacific Ocean variety (live & processed)
PLATE II	3. *Actinopyga echinites* Indian Ocean variety (live)
	4. *Actinopyga lecanora* (live & processed)
PLATE III	5. *Actinopyga mauritiana* Pacific Ocean variety (live & processed)
	6. *Actinopyga miliaris* (live & processed)
PLATE IV	7. *Actinopyga palauensis* (live & processed)
	8. *Actinopyga spinea* (live & processed)
PLATE V	9. *Actinopyga* sp. affn. *flammea* (live & processed)
	10. *Bohadschia argus* (live & processed)
PLATE VI	11. *Bohadschia atra* (live & processed)
	12. *Bohadschia marmorata* (live & processed)
PLATE VII	13. *Bohadschia subrubra* (live & processed)
	14. *Bohadschia vitiensis* (live & processed)
PLATE VIII	15. *Pearsonothuria graeffei* (live & processed)
	16. *Holothuria arenicola* (live & processed)
PLATE IX	17. *Holothuria atra* (live & processed)
	18. *Holothuria cinerascens* (live & processed)
PLATE X	19. *Holothuria coluber* (live & processed)
	20. *Holothuria edulis* (live & processed)
PLATE XI	21. *Holothuria flavomaculata* (live)
	22. *Holothuria fuscocinerea* (live & processed)
PLATE XII	23. *Holothuria fuscogilva* (live & processed)
	24. *Holothuria fuscopunctata* (live & processed)
PLATE XIII	25. *Holothuria hilla* (live)
	26. *Holothuria impatiens* (live)
PLATE XIV	27. *Holothuria kefersteini* (live & processed)
	28. *Holothuria lessoni* blotchy variant (live & processed)
PLATE XV	29. *Holothuria lessoni* beige variant (live)
	30. *Holothuria lessoni* black variant (live)
	31. *Holothuria leucospilota* (live)
PLATE XVI	32. *Holothuria mexicana* (live & processed)
	33. *Holothuria nobilis* (live & processed)
PLATE XVII	34. *Holothuria notabilis* (live & processed)
	35. *Holothuria* sp. (type 'Pentard') (live & processed)

page 12

← live

photo by: J.J. Alvarado

1. *Actinopyga agassizii*

Processed photo not available.
Courtesy submissions are welcome.

← processed

page 14

← live
Pacific Ocean variety
photo by: S.W. Purcell

2. *Actinopyga echinites*

← processed
Pacific Ocean variety
photo by: S.W. Purcell

page 14

← live
Indian Ocean variety
photo by: P. Bourjon

3. *Actinopyga echinites*

← processed
Indian Ocean variety

page 16

← live

photo by: S.W. Purcell

4. *Actinopyga lecanora*

← processed

photo by: S.W. Purcell

page 18

← live
Pacific Ocean variety
photo by: S.W. Purcell

5. *Actinopyga mauritiana*

← processed
Pacific Ocean variety
photo by: S.W. Purcell

page 20

← live

photo by: S.W. Purcell

6. *Actinopyga miliaris*

← processed

photo by: S.W. Purcell

page 22

← live

photo by: S.W. Purcell

7. *Actinopyga palauensis*

← processed

photo by: S.W. Purcell

page 24

← live

photo by: S.W. Purcell

8. *Actinopyga spinea*

← processed

photo by: S.W. Purcell

page 26

← live

photo by: G. Paulay

9. *Actinopyga* **sp. affn.** *flammea*

← processed

photo by: S.W. Purcell

page 28

← live

photo by: S.W. Purcell

10. *Bohadschia argus*

← processed

photo by: S.W. Purcell

page 30

←— live

photo by: P. Frouin

11. *Bohadschia atra*

←— processed

photo by: C. Conand

page 32

←— live

photo by: S.W. Purcell

12. *Bohadschia marmorata*

←— processed

photo by: E. Aubry SPC

page 34

← live

photo by: P. Laboute

13. *Bohadschia subrubra*

← processed

photo by: H. Eriksson

page 36

← live

photo by: S.W. Purcell

14. *Bohadschia vitiensis*

← processed

photo by: S.W. Purcell

page 38

← live

photo by: S.W. Purcell

15. *Pearsonothuria graeffei*

← processed

photo by: J. Akamine

page 40

← live

photo by: A.M. Kerr

16. *Holothuria arenicola*

← processed

photo by: S.W. Purcell

page 42

← live

photo by: S.W. Purcell

17. *Holothuria atra*

← processed

photo by: S.W. Purcell

page 44

← live

photo by: P. Bourjon

18. *Holothuria cinerascens*

← processed

photo by: C. Conand

page 46

← live

photo by: S.W. Purcell

19. *Holothuria coluber*

← processed

photo by: J. Akamine

page 48

← live

photo by: S.W. Purcell

20. *Holothuria edulis*

← processed

photo by: S.W. Purcell

page 50

← live

photo by: S.W. Purcell

21. *Holothuria flavomaculata*

Processed photo not available.
Courtesy submissions are welcome.

← processed

page 52

← live

photo by: G. Edgar

22. *Holothuria fuscocinerea*

← processed

photo by: J. Akamine

page 54

← live

photo by: S.W. Purcell

23. *Holothuria fuscogilva*

← processed

photo by: S.W. Purcell

page 56

← live

photo by: S.W. Purcell

24. *Holothuria fuscopunctata*

← processed

photo by: S.W. Purcell

page 58

← live

photo by: S.W. Purcell

25. *Holothuria hilla*

Processed photo not available.
Courtesy submissions are welcome.

← processed

page 60

← live

photo by: S. Ribes

26. *Holothuria impatiens*

Processed photo not available.
Courtesy submissions are welcome.

← processed

page 62

← live

photo by: G. Edgar

27. *Holothuria kefersteini*

← processed

photo by: S.W. Purcell

page 64

← live
blotchy variant
photo by: S.W. Purcell

28. *Holothuria lessoni*

← processed

photo by: B. Giraspi

page 64

← live
beige variant
photo by: S.W. Purcell

29-30. *Holothuria lessoni*

← live
black variant
photo by: S.W. Purcell

page 66

← live

photo by: S.W. Purcell

31. *Holothuria leucospilota*

Processed photo not available.
Courtesy submissions are welcome.

← processed

page 68

← live

photo by: SIMAC-INVEMAR

32. *Holothuria mexicana*

← processed

photo by: F.A. Solís-Marín

page 70

← live

photo by: R. Aumeeruddy

33. *Holothuria nobilis*

← processed

photo by: S.W. Purcell

page 72

← live

photo by: IH-SM-WIOMSA

34. *Holothuria notabilis*

← processed

photo by: IH-SM-WIOMSA

page 74

← live

photo by: R Aumeeruddy

35. *Holothuria* sp. (type 'Pentard')

← processed

photo by: C. Conand

page 76

←——— live

photo by: www.noaa.gov

36. *Holothuria pardalis*

Processed photo not available.
Courtesy submissions are welcome.

←——— processed

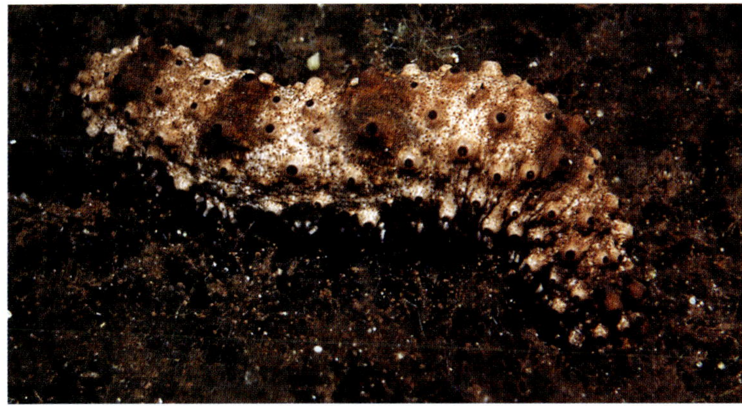

page 78

←——— live

photo by: K. Stender

37. *Holothuria pervicax*

Processed photo not available.
Courtesy submissions are welcome.

←——— processed

page 80

← live
Pacific Ocean variety
photo by: S.W. Purcell

38. *Holothuria scabra*

← processed
Pacific Ocean variety
photo by: S.W. Purcell

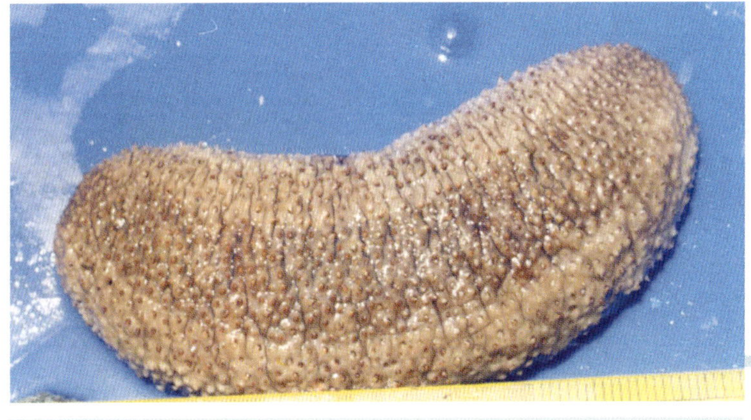

page 82

← live

photo by: P.S. Asha

39. *Holothuria spinifera*

← processed

photo by: D.B. James

page 84

← live

photo by: S.W. Purcell

40. *Holothuria whitmaei*

← processed

photo by: S.W. Purcell

page 86

← live

photo by: A. Semenov

41. *Apostichopus japonicus*

← processed

photo by: S.W. Purcell

page 88

← live

photo by: J.M. Watanabe

42. *Apostichopus parvimensis*

← processed

photo by: J. Akamine

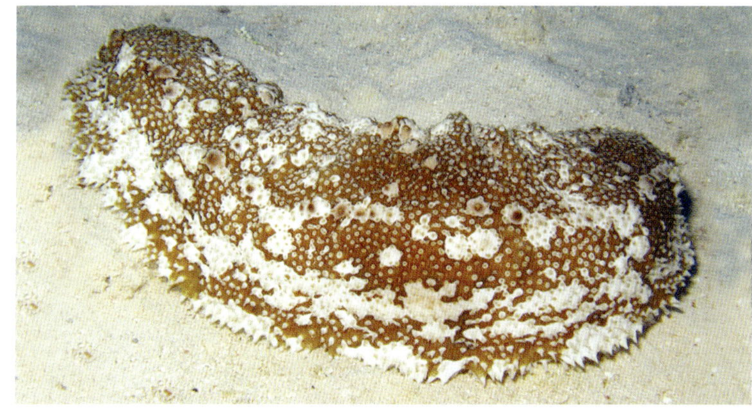

page 90

← live

photo by: F. Charpin

43. *Astichopus multifidus*

Processed photo not available.
Courtesy submissions are welcome.

← processed

photo by: S.W. Purcell

page 92

←—— live

photo by: K. Clemments

44. *Australostichopus mollis*

←—— processed

photo by: L. Zamora

page 94

←—— live

photo by: E. Ortiz

45. *Isostichopus badionotus*

←—— processed

photo by: S.W. Purcell

page 96

← live

photo by: S.W. Purcell

46. *Isostichopus fuscus*

← processed

photo by: S.W. Purcell

page 98

← live

photo by: J.M. Watanabe

47. *Parastichopus californicus*

← processed

photo by: J. Akamine

page 100

← live

photo by: S.W. Purcell

48. *Stichopus chloronotus*

← processed

photo by: S.W. Purcell

page 102

← live

photo by: S.W. Purcell

49. *Stichopus herrmanni*

← processed

photo by: S.W. Purcell

page 104

←——— live

photo by: G. Paulay

50. *Stichopus horrens*

←——— processed

photo by: S.W. Purcell

page 106

←——— live

photo by: S.W. Purcell

51. *Stichopus monotuberculatus*

←——— processed

photo by: S.W. Purcell

page 108

← live

photo by: S.W. Purcell

52. *Stichopus naso*

← processed

photo by: C. Dissanayake

page 110

← live

photo by: S.M. Wolkenhauer

53. *Stichopus ocellatus*

← processed

photo by: L.B. Concepcion

page 112

← live

photo by: D. VandenSpiegel

54. *Stichopus pseudohorrens*

Processed photo not available.
Courtesy submissions are welcome.

← processed

page 114

← live

photo by: S.W. Purcell

55. *Stichopus vastus*

← processed

photo by: S.W. Purcell

page 116

← live

photo by: S.W. Purcell

56. *Thelenota ananas*

← processed

photo by: J. Akamine

page 118

← live

photo by: S.W. Purcell

57. *Thelenota anax*

← processed

photo by: S.W. Purcell

page 120

← live

photo by: K. Friedman

58. *Thelenota rubralineata*

← processed

photo by: L.B. Concepcion

page 122

← live

photo by: L. Amaro-Rojas

59. *Athyonidium chilensis*

← processed

photo by: C. Guisado

page 124

← live

photo by: J.F. Hamel &
A. Mercier

60. *Cucumaria frondosa*

← processed

photo by: S.W. Purcell

page 126

← live

photo by: N. Sanamyan

61. *Cucumaria japonica*

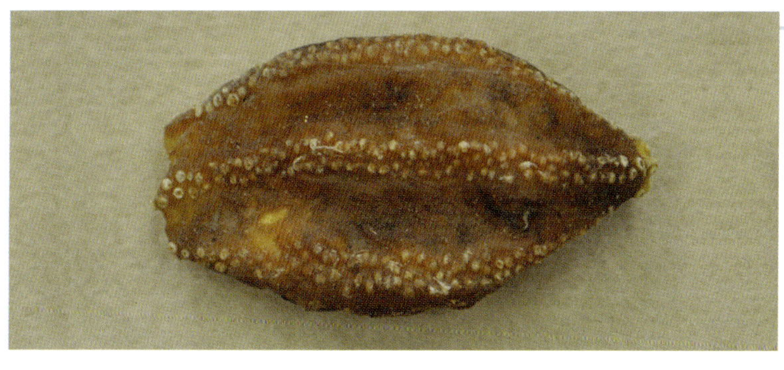

← processed

photo by: J. Akamine